CAMBRIDGE PHYSICAL SERIES.

GENERAL EDITORS:—F. H. NEVILLE, M.A., F.R.S.
AND W. C. D. WHETHAM, M.A., F.R.S.

AIR CURRENTS

AND THE

LAWS OF VENTILATION

AIR CURRENTS

AND THE

LAWS OF VENTILATION

LECTURES ON THE PHYSICS
OF THE VENTILATION OF BUILDINGS
DELIVERED IN THE UNIVERSITY OF CAMBRIDGE
IN THE LENT TERM, 1903

by

W. N. SHAW, Sc.D., F.R.S.
Honorary Fellow of Emmanuel College
Director of the Meteorological Office

Cambridge
at the University Press
1907

CAMBRIDGE
UNIVERSITY PRESS

University Printing House, Cambridge CB2 8Bs, United Kingdom

Published in the United States of America by Cambridge University Press, New York

Cambridge University Press is part of the University of Cambridge.

It furthers the University's mission by disseminating knowledge in the pursuit of education, learning and research at the highest international levels of excellence.

www.cambridge.org
Information on this title: www.cambridge.org/9781107690608

© Cambridge University Press 1907

First published 1907
First paperback edition 2014

A catalogue record for this publication is available from the British Library

ISBN 978-1-107-69060-8 Paperback

PREFACE.

AMONG the duties laid upon me by the tenure of a fellowship at my old College from 1900 till 1906 was the delivery each year of a short course of lectures before the University, upon some subject connected with the Physics of the Atmosphere. This little book represents my endeavour to discharge that duty in the year 1903.

For many years I have taken a practical interest in the physical aspects of the general problem of ventilation. When I was a member of the Museums and Lecture Rooms Syndicate, I tried to meet some complaints about the state of the air in the Biological Lecture Room by making use of the furnace draught of the boilers in a neighbouring stoke-hole and communicated an account of the arrangement to the Philosophical Society. The experiment was successful enough from the physical point of view but not from the biological, for the stoker found that, when it was working, the furnace room became uncomfortably warm and any system of ventilation which disregards the comfort of the stoker has serious disadvantages to contend with.

The experiment, however, led to my undertaking the article on "Ventilation and Warming" for Stevenson and Murphy's *Hygiene* and that again to the inspection of the ventilation of Metropolitan Poor Law Schools for the Local Government Board. Since then I have been appealed to for advice as regards ventilation from many directions.

In the course of the preparation of the article for Stevenson and Murphy's *Hygiene*, I came across a small book by M. Murgue, a Belgian engineer, translated by A. L. Steavenson, upon the application of centrifugal ventilating machines to the ventilation of mines. In that book the idea of replacing or representing the pneumatic resistance of a complicated channel by its equivalent thin-plate-orifice is developed and the law of relation of head to flow, which may be regarded as the basis of all ventilation arrangements, is clearly set out. I was struck with the analogy which the relation presented to the distribution of electrical currents in a net-work of conductors and the possibility thus afforded of reducing the problems of ventilation from the chaos of mere qualitative suggestion, by means of diagrams with arrows indicating the paths in which it was desired that air should move, to something like the order of numerical relationship, and I discussed the matter in a short communication to the British Association at Leeds in 1890. Lord Kelvin in commenting upon the paper pointed out the curious reversal of relations, in that the knowledge of electricity, which in his earlier days was illustrated by the more familiar case of pneumatics, should have become so widespread that it could now be appealed to for illustration of the more complicated relations of pneumatics.

I carried the idea of the electrical analogy further by making out of two boxes, communicating with a third in which there was an improvised chimney shaft, a pneumatic analogue of the Wheatstone quadrilateral, which verified the accuracy of the pneumatic laws and their application to the division of air currents in the most conclusive manner. With the assistance of Mr R. S. Cole, a number of most interesting measurements were made with this apparatus. Indeed the use of the apparatus was carried on with complete success, so far as its size would permit. It was only a laboratory model, and our experiments were necessarily confined to small openings or tubes of about an inch diameter. I feel sure that what is most needed at the present day for the development of an accurate knowledge of the management of pneumatic currents is the carrying out of similiar experiments on the larger and more practical scale appropriate to the Municipal Technical School, instead of on the restricted scale of a Physical Laboratory. The information which could thus be obtained as to the characteristic curves of fans and the effect of divers small variations of condition upon the flow through orifices and tubes would, I feel sure, be at once interesting, instructive and stimulating. It is a matter of great regret that, so far as I know, the real study of pneumatics finds no place in our great technical establishments.

Ventilation, like the weather, is a subject of universal interest and gives rise to the widest differences of opinion. One man's fresh air is another man's draught, and the most difficult part of the ventilation problem is to reconcile the interests of both these classes of persons.

In the practical attempts to solve the problem, too little attention has, I think, been paid to the laws of physics. Hygienic chemists have told us what they consider to be the permissible limit of respirable impurity in air, and various writers have expressed opinions as to what degree the thermometer should mark to keep up healthy conditions; but neither of these important facts suffices to tell us how the supply of air designed to wash away the respirable impurity and maintain the requisite temperature may be expected to fulfil its very difficult task by its journey from an inlet to an outlet.

Take an example: I suppose the greatest social enemy from the ventilation point of view is the individual who sneezes; the person who merely breathes is bad enough, the person who sneezes adds a special difficulty because he distributes a cloud of fine particles which may or may not be deleterious. Imagine such an enemy in a remote corner of a large room and consider what must be done to wash away the respirable impurity. Some benevolent authority may have decided perhaps that 1000 cubic feet of air per hour is sufficient for the work, because no signs of exceptional discomfort or illness have manifested themselves when that amount of air has been supplied, on the average. So, every second, a quarter of a cubic foot of air is duly provided for our enemy and delivered for him through an opening in the wall, which some say should be high up, others low down. At the same time the quarter cubic foot which was supplied for him some fifteen minutes previously is called for at another opening in another wall, which again some say should be low down and others high up. But how is our enemy with the sneezing cold to be sure of getting his quarter cubic foot, and how shall he be sure of giving it up when he has used it in such a way that it shall reach the proper opening for its removal? As a matter of fact the distribution of the air supply between the occupants of a room is a matter of scrambling between them, with the aid of the convection current which each individual causes; and in the course of the scramble there is much scope for the exercise of various physical laws, that are not much regarded in deciding the general practice of ventilation. Certain it is that, with a single opening in one wall for delivery and a single opening in another for extraction, if every occupant of a school class room does get his fair share and use of the fresh air supply, it is a result that one could not anticipate, and it is well worth consideration from the point of view of experimental physics.

Accordingly, in the middle section of this book I have laid some stress upon the physics of the ventilated space. The conclusion which I draw myself from the consideration of these questions is that the requirements of the physical process indicate a large air supply as a necessity for a solution of the difficult problem of ventilation without draughts. This may seem paradoxical, and it is possible to put it in still more obviously paradoxical form. If complaint is made of draught the proper remedy may be to supply more air—perhaps a little warmer—not less air a little colder or of the same temperature. The explanation of the paradox is that air is required to wash away the surplus heat as well as the surplus impurity, that the draught may be due to the united effort, in the way of convection, of the assembly of people and that the effects of convection are less if there is more air with which to work.

This volume is in a way my last will and testament in respect of ventilation. Within the last few months I have been asked for advice with regard to the ventilation of the new buildings of the University of Wales at Cardiff, the House of Commons, the meeting room of the Royal Society, the examination rooms of the University of Cambridge, with various other special cases, and generally about the ventilation of schools by a committee of the British Association. From what follows it will be understood that each one of these questions suggests the determination of the distribution of currents in a net-work of electrical conductors under the action of various electromotive forces. The answer is really represented by the solution of a large number of simultaneous simple equations. I do not pretend that in the ventilation problem the equations would all be expressed or worked out arithmetically. If they were, many of the coefficients would have to be guessed. But all the same the problem depends on the solution of the simultaneous equations. All the various contingencies have to be taken into account, some disregarded as unimportant, others specially considered, and an opinion arrived at which takes some sort of account of each of them. In the same way, if we had a series of simultaneous equations, no one would, I think, embark upon a rigid arithmetical solution if he could simplify his relations by more or less justifiable approximations.

Now to get the conditions thus represented into one's mind

so completely that one can form an effective opinion, by this process of neglecting the inconsiderable and appreciating the important, is a work that requires a good deal of undisturbed attention. The requirements of the study of the processes of ventilation on the larger scale, as they are exhibited in the weather, no longer permit me to work out these interesting but complicated problems. I have therefore written down in this volume all that I know, and perhaps added a good deal that I have guessed, that will enable anyone, who is so disposed, to take up the numerical consideration of ventilation questions on the basis of the electrical analogy and with due regard to the physical conditions of the distribution of air.

I could wish, as I have already hinted, that the experimental study of such problems could be regarded as a part of the duty of schools of technical science. It can scarcely be doubted that the subject is of sufficient interest and importance.

Mr Whetham has been good enough to read the proofs and has rendered thereby timely assistance to a work which could only be taken up at long intervals.

My thanks are also due to the President and Council of the Royal Society for permission to use the block from which Figure 6 was prepared, to the Royal Sanitary Institute for Figures 4 and 5, to Messrs J. and A. Churchill for Figures 13, 14, 15, 16, and to Mr W. H. Hayles of the Cavendish Laboratory for photographs for a number of the illustrations, as well as for practical assistance during the lectures.

W. N. SHAW.

April 3, 1907.

CONTENTS.

I.

LAWS OF FLOW IN AIR CIRCUITS AND THEIR VERIFICATION.

THE process of ventilation consists in the passage of air into the space to be ventilated and out of it again, and the distribution of the air during its passage through the ventilated space. We may treat these two parts of the process separately, and consider first the conditions necessary for maintaining steady currents of air through the inlets and outlets, postponing to a subsequent section the consideration of the distribution of air in the ventilated space.

Definition of Air Circuit.

I propose to call a complete arrangement for the flow of air an air circuit. It is obvious that the circuit must begin and end in the external air; it can be regarded as composed of an inlet channel and an outlet channel between the outside air and the ventilated space and some motive power, natural or artificial, which maintains the flow. The circuit may be simple, i.e. it may have a single inlet and outlet and a single motive power, or it may be complex by the reduplication of any one of its elements. It may have several inlets or several outlets and each inlet or outlet may be provided, intentionally or unintentionally, with independent sources of motive power. I will consider first the laws applicable to the flow in air circuits, and primarily in a simple circuit.

Law I. Continuity of Flow.

The first and most obvious law may be styled the law of continuity of flow. It expresses the simple fact that in any ventilated space air cannot go on accumulating, nor can any limited space be drawn upon for an unlimited supply of air. With the exceptions hereafter mentioned, to replace whatever air is taken out of a room by the outlet an equal quantity must be supplied by inlets and *vice versa,* or for a simple circuit the amount of air passing any two cross-sections of the circuit must be the same wherever the cross-sections may be. For this law to hold strictly the amount of air should be estimated by the *weight* crossing a section in a definite time. The volume may be different for different parts of the circuit if they are at different temperatures or substantially different pressures. It is however convenient to estimate the flow of air by the volume of air passing, and not by the weight, because the volume is easily deduced from the speed of delivery*. We have therefore to decide at the outset whether we shall adopt the accurate method of estimation of flow by weight, although it involves additional experimental measures, or use the volume measure with the uncertainty attaching thereto on account of the variation of density.

It may be noticed however that except when air is passing up a chimney or some other strongly heated flue the effect of temperature on the density of the air is small. For example, a temperature change of 30° C. or 54° F. say from an outside temperature of 14° F. to an inside temperature of 68° F., which would be a very large temperature difference for a ventilation circuit, would only affect the readings by about 10 per cent. and, in the present state of pneumatic measurements, would not be an error of extraordinary magnitude compared with other experimental errors. The modifications in the calculations which would be necessary to take account of the variations of density could be introduced if required. In the present position of the subject of ventilation, which is only just becoming amenable to quantitative measurements, any such modification is an

* See Note A, p. 89.

unnecessary refinement. Accordingly, unless we are dealing with hot flues we shall take the air density of a circuit as constant so far as temperature is concerned and represent it by ·08 lb. per cubic ft. A cubic foot of chemically dry air actually weighs ·08 lb. when its temperature is 34·5° F. and its pressure that of 29·9 inches of mercury. It does not actually weigh that amount under other conditions.

Weight of air required for ventilation.

This estimation of the density of air will serve to show what large weights of air have to be moved in any ventilation system of considerable magnitude. Upon the usual computation every individual in this room requires about 240 lbs. of air for his use during the lecture or more than $\frac{1}{10}$ of a ton; 10 people want more than a ton; 100 people more than 10 tons. If you wish to know how this room could be ventilated for its full capacity (300 persons) you have to consider how 30 tons of air can be brought into the room and taken out again in the course of the hour. For proper ventilation a church with a congregation of 500 persons requires a supply of 50 tons of air during the service*.

Further consideration of the law of continuity.

I have passed over pressure differences as affecting the density of air in its course. Consider what they are. The pressure difference of one inch of water is a very substantial one for ventilating systems in ordinary buildings; that is only the $\frac{1}{400}$ part of the atmospheric pressure and therefore quite out of consideration as compared with temperature changes.

I have referred to exceptions to the law of continuity. These occur only when the flow is not steady, i.e. when the temperature or pressure is fluctuating in different parts of the circuit. A ventilated space will hold more air at a lower temperature or higher pressure, and the air pressure in a ventilated space has to be raised or lowered by increasing or diminishing the quantity of air in the space before a state of things is reached under which as much passes out as comes in.

* The computation is based on the allowance of 3000 cubic ft. per hour per person. The suitability of this amount is considered later, page 56.

With very large spaces, differences in the quantity of air contained in the space under different conditions are important. The most striking domestic results of these variations from the law of continuity are the carrying of dust to the interior of drawers, cupboards and other nearly closed spaces by the "breathing" that corresponds with alternations of high and low pressure or temperature. Similar processes on a large scale have been held to account for explosions in coal mines, and may perhaps, by the "breathing" of porous soil, account for some appreciable modification of the hygienic state of the air under certain weather conditions.

When combustion takes place in the course of the process of ventilation, the law of continuity is again interfered with. But again, the effect upon the volume of air delivered is not important in the circumstances usually occurring; four-fifths of the air which maintains the combustion takes no part in it, and the volumes of the resulting products of combustion do not differ seriously from that of the original oxygen; hence the change of volume resulting from combustion is not much greater than that which is represented by the change of temperature incidental to the combustion.

The law of continuity of flow is now generally accepted in theory. It is still frequently ignored in practice. Some people are still under the impression that if you make provision for taking air out you have done all that ventilation requires, or similarly if you make provision for bringing air in. It is perfectly true that, whatever provision you make, just as much air will go out as comes in, and just as much will come in as goes out, but not always in exactly the way that you expect, unless you recognise beforehand the way in which the law of continuity of flow will apply itself to your particular arrangement.

We may express the law of continuity of flow algebraically as follows: If V be the flow in cubic feet per second and ρ the density of air, $V\rho$ is constant for all cross-sections of a simple circuit, and, since ρ is generally constant within the limits of experiment, V is also constant within similar limits for all cross-sections.

When circuits are complex and there are many inlets and

outlets, the statement of the law of continuity requires modification to the effect that the total amount of air passing through all the inlets for the time being is equal to the total amount of air passing through all the outlets. A similar understanding as to the steadiness of the flow must be presupposed. With complex circuits steadiness of flow is not so easily maintained as with a simple one.

Law II. The relation of the Flow in an air circuit to the Head, or Aeromotive force, producing it.

In order to maintain a continuous flow in a ventilation circuit energy must be continuously expended. It is useless to have openings by which air may enter or leave the space which it is intended to ventilate unless there is some motive power to drive air through the openings. For most enclosed spaces if openings are provided, there usually exists some motive power due to wind or heated air which has been called "natural," as distinguished from the artificial motive power which may be produced by flues, or chimney-stacks with fires specially maintained, or by ventilating fans or blowing engines supplied with mechanical power for the express purpose of maintaining a flow in a circuit. Perhaps the more effective distinction between the so-called "natural" and other sources of power for ventilation lies in the fact that the natural motive power, whether due to wind or heated air, costs nothing to maintain it for ventilation purposes, the heat used being a by-product of some other process, while a separately maintained fire or the mechanical agent for driving a ventilating fan is a source of expense as well as of power. As long as we have to deal with the mere existence of a flow in the circuit without having to base any serious calculations upon its direction or its magnitude in relation to other quantities involved, the motive power may well be left to nature; but when a more or less precise adjustment of the flow is required, the effectiveness of the motive power must be definitely estimated and provided for. It is an easy matter to arrange a dwelling space so incompletely closed as to allow an automatic flow of air sufficient for a single person, but as the number of persons increases, satisfactory arrangements depending upon automatic flow become increasingly difficult.

Motive power for ventilation may be derived from wind in two ways, by blowing directly upon an opening and by blowing across an opening under suitable conditions. The direction of flow may be said to be opposite in the two cases. In the first case the air is driven into the aperture by the pressure of the wind; in the second the air is extracted from the aperture. Ventilation by wind-force in any particular arrangement may result from a combination of these two effects. Other available means of causing flow in an air circuit are, first, the heating of air in some considerable vertical stretch of the circuit and, secondly, air-fans and blowing engines. Whatever the motive power may be, if the flow is to be maintained, the working agent must be in continuous operation, the wind must go on blowing, the fire that warms the flue must be maintained, the engine driving the fan must be kept at work. All this means continuous expenditure of energy, continuous work, whether it is separately provided and paid for or not. Moreover if the ventilation is to be steady and maintained, if the flow is to be definite in amount, the agent, whatever it is, must work steadily and the work that it does in any given time must be equally definite in amount.

We thus obtain an idea of a constant performance of work, a constant expenditure of energy, resulting in a constant flow of a definite quantity of air through an air circuit. We may regard the air circuit as offering a sort of resistance to the flow of air in consequence of its shape and size, and regard the motive power necessary to maintain the flow as spent in overcoming the " resistance " of the circuit. In this respect there is an analogy between the maintenance of flow in an air circuit and the maintenance of an electric current in a circuit of electrical conductors. Mechanical power is necessary to overcome electrical resistance and maintain electric current; mechanical power is similarly necessary to overcome pneumatic resistance and maintain a flow of air.

We require a means of estimating numerically the power required to maintain an air-flow of given magnitude in an air circuit. I shall take the work required to get one pound of air through the circuit as the measure of motive power or "aero-motive force," just as the work required to get one unit of electricity through an electric circuit is the measure of the electromotive force.

It is no doubt a tribute to the extension of a general knowledge of electricity within the last 50 years, as Lord Kelvin pointed out, that the laws of electrical circuits should be referred to now to explain and illustrate pneumatic laws when 50 years ago the reverse would have been the case. The application of mechanical power of measured amount to the maintenance of electrical currents is at this day much more familiar as an example of the use of power than the corresponding process for pneumatics.

The relation of the aeromotive force as here defined to the horse power required to maintain a constant flow of air is easily obtained. If V is the flow in cubic feet per second, maintained by an aeromotive force H, and ρ is the density of the transmitted air in pounds per cubic foot, $V\rho$ is the quantity of air passing per second in the steady flow, and the number of foot-pounds of work required to maintain the flow is by the definition of H, $HV\rho$. The horse power P required is consequently $HV\rho/550$. This equation enables us to calculate from the aeromotive force the horse power necessary to maintain a measured flow, or *vice versa*. For example the aeromotive force in the flue of an ordinary chimney may be assessed at 6 foot-lbs. per pound of air*, the flow in an ordinary chimney may be taken as 10,000 cubic feet of air per hour or 2·8 cubic feet of air per second: the power utilised for working such a chimney is therefore at the rate of 6 × 2·8 × ·08/550, i.e. about one four-hundredth of a horse power. This power is derived from the combustion of the fuel, by which the temperature in the flue is maintained above that of the air outside the flue.

Aeromotive force may also be expressed as "head" or pressure difference, and this is in many ways the simplest representation of the work done in carrying a pound of air through a circuit. From the physical dimensions of the quantities it is clear that the magnitude "foot-pounds per pound of air" is a length, or height, and is not directly dependent on any unit of mass,

* See Note B, p. 91.

though it depends upon the value of gravity. It is equivalent to the height to which a pound of air would be lifted by the work which would carry it through the air circuit, or the height through which it must descend in order to overcome the resistance offered to the passage of the same quantity through the circuit. It is thus equal to the " head " of air which must be maintained in order to maintain the flow. The same head could be maintained by a column of water which would produce the same pressure as the air column, and thus aeromotive force, whencesoever it is derived, may be expressed as the "head" due to a column of air, or the equivalent column of water.

It therefore becomes possible to compare the different agencies for producing flow of air by specifying the head, or aeromotive force, produced by each agency under specified conditions. I computed a table exhibiting the conditions necessary for the production of specified aeromotive forces ranging from ·67 unit to 670 units for the article on " Warming and Ventilation " in Stevenson and Murphy's *Hygiene*, vol. I. p. 84. Such a table involves a considerable number of experimental determinations which are all more or less uncertain and are subject to revision as more accurate pneumatic experiments become available ; but the table is sufficient to give an indication of the relative magnitudes of the head or aeromotive force obtainable from different sources. A modification of the table based on the same calculations is given here, and in Note B, p. 91, will be found an account of some of the experimental data upon which the calculations are based. In the table here given, the aeromotive force in pneumatic units as defined above is taken as the argument, and the equivalent in terms of a column of water, and the pressure in pounds per square foot are computed.

The second law of ventilation deals with the relation between the flow maintained in the circuit and the " head " or aeromotive force which maintains it. If we consider the way in which power is spent, or work done, in carrying the air from the outside of a space through an inlet, which may be a short channel or a long one, through the ventilated space and through the outlet, it will be evident that the power has to produce the velocity in the inlet, to alter the velocity as the inlet varies in width, to reproduce velocity in the outlet after it has been practically

Table showing the amount of aeromotive force or "head" produced by various agencies.

Aeromotive force in pneumatic units ("Head in 'feet of air'")	Equivalent water column	Equivalent pressure	Horse power expended per "unit of ventilation" (1 cubic ft. of air per second)	Velocity of wind for direct impact	Velocity of wind passing over the mouth of a duct	Velocity of tips of vanes of centrifugal fans		Velocity of tips of vanes of helicoidal fans		Internal temperature of a 40 ft. shaft External air assumed to be at 41° F.
"Feet of air"	Inches	lbs. weight per sq. ft.	H. P. per unit	Feet per sec.	Feet per sec.	Feet per sec. Worst	Best	Feet per sec. Worst	Best	° F.
1	·015	·08	·00014	9·0	7·9	17·7	6·17	25·0	12·5	54
2	·03	·16	·00028	12·6	11·3	25·3	9·55	35·8	17·9	63
3	·05	·23	·00042	15·5	13·8	31·0	11·7	43·8	22·0	78
4	·06	·31	·00056	17·0	16·4	35·9	13·5	50·6	25·3	91
5	·08	·39	·00070	21·6	17·9	40·0	15·1	56·7	28·4	103
6	·09	·47	·00084	23·3	19·8	43·9	16·5	62·0	31·0	115
7	·11	·55	·00098	25·9	21·1	47·3	17·8	66·9	33·4	128
8	·12	·62	·00112	27·4	22·4	50·1	18·9	70·8	35·4	140
9	·14	·70	·00126	39·0	23·7	53·0	20·0	74·6	37·5	153
10	·15	·78	·00140	54·9	25·0	55·9	21·1	78·6	39·5	165
20	·30	1·56	·00280	61·1	35·8	80·1	30·2	114	56·6	291
30	·45	2·34	·00420	84·9	43·7	94·1	37·0	138	69·3	
40	·60	3·12	·00560		50·6	113	42·8	160	80·1	
50	·75	3·90	·00700		56·4	123	47·9	180	89·5	
100	1·50	7·80	·01400		78·9	177	66·8	248	125	
200	3·00	15·60	·02800			253	95·7	358	179	
300	4·50	23·40	·04200			309	117	438	219	
400	6·00	31·20	·05600			359	135	506	253	
500	7·50	39·00	·07000			400	151	566	284	

lost in the ventilated space, to alter its direction in accordance with the bends of the channels, and to overcome friction in any considerable straight length. Now it is well-known that in a very long uniform channel part of the expenditure of energy depends upon friction proportional to the velocity, whereas the production of velocity, or kinetic energy, will require power proportional to the square of the velocity produced. Thus if we desire minute accuracy we should have to assume the law connecting the "head" and flow to be of a complex character such as

$$H = RV^2 + R'V,$$

where R' gives the power spent in overcoming the frictional resistance of unit flow, and R the power spent in generating velocity and providing for all other demands for energy which depend upon the square of the flow.

I wish to point out that in all ordinary arrangements for ventilation that part of the expenditure of energy which depends on V, the first power of the flow, is negligible in comparison with the part depending on V^2—in other words R' is negligible compared with R, and for all practical purposes the law of relation between head and flow in ventilation circuits is $H = RV^2$.

Direct evidence for Law II.

For direct evidence in support of this statement I must refer to Péclet's *Traité de la Chaleur*, vol. III. In that work the items of expenditure of energy in the various circumstances through which air has to pass, on its way through a ventilation circuit, are analysed and separately estimated. The production of velocity, the loss of energy due to passing an angle, to rounding a bend, to a sudden diminution in the area of a channel, to a sudden enlargement of the area of a channel; all these are found experimentally to be proportional to the square of the flow. There remains only the smooth-skin friction which gives a term proportional to the flow itself, and skin friction is of very small importance in practical ventilation. The expenditure of energy arises from the production of kinetic energy and the eddies due to the turbulent motion of the air passing obstacles of various kinds.

Indirect evidence for the truth of the law.

The theoretical investigation of the expenditure of energy from the various causes specified is beyond my scope, and it is moreover unnecessary; the most effective method of verifying a physical law is to determine the practical consequences which follow from its acceptance and to compare those consequences with the results of experience. I propose to proceed in a similar manner with the law under consideration, to consider first the consequences that follow from the assumption that the law $H = RV^2$ holds for any simple ventilation circuit and subsequently to compare the conclusion with experimental results for certain circuits.

Electrical Analogy. Standard of Resistance.

The law of relation between head and flow of air $H = RV^2$ may be compared with Ohm's law for the flow of electricity in an electric circuit, viz. $E = RC$, although the flow enters in the second power in the pneumatic relation, whereas electric current appears in the first power in the electrical one. Recognising that difference, R may be called the pneumatic resistance of the circuit, and may be defined as the head required to produce unit flow in the circuit, or the energy required per pound of air to maintain a flow of one cubic foot per second. One cubic foot per second is on the ordinary basis of computation a liberal allowance of air for one person, and thus the resistance of a ventilation circuit may be defined provisionally as the head required in any given ventilation circuit to give ventilation for one person.

For practical electrical purposes a standard of resistance has been selected, the ohm, as represented by the Board of Trade Standard. It may be of interest to consider what the corresponding pneumatic standard would be. It would be a channel through which unit head would maintain unit flow. If we take a head, or aeromotive force, of one foot of air as corresponding with the volt, and the flow of one cubic foot of air per second as corresponding with the ampère, the resistance of an air channel which requires a head of one foot to maintain a flow of one cubic foot per second will represent the standard of pneumatic resistance corresponding with the ohm, and all other resistances can be expressed subsequently in terms of this standard.

For electric current a long wire represents the most typical form of resistance; for an air circuit an aperture in a very thin plate is selected because the law of pneumatic resistance, $H = RV^2$, is strictly obeyed in this case.

I have shown elsewhere* that if α is the area of an aperture in a thin plate expressed in square feet, the resistance of the aperture to the flow of air is $\dfrac{1}{27\alpha^2}$. Whence it follows that the unit or standard resistance is that of a thin plate aperture of area α, where $\alpha^2 = \frac{1}{27}$, or $\alpha = \cdot189$ square foot.

If the aperture is circular and of diameter d,

$$\pi d^2/4 = \cdot189 \quad \text{or} \quad d = \cdot498 \text{ ft.}$$

The circular aperture has thus a diameter of almost exactly 6 inches.

It follows thus that a circular aperture of 6 inches diameter represents very approximately the practical standard of pneumatic resistance. Through such an aperture unit head will cause unit of ventilation flow, the flow necessary for a liberal allowance of ventilation for a single person. Unit head, a foot of air, corresponds approximately with a pressure difference of 1/800 ft. of water or ·015 inch.

We thus obtain the fundamental statement for the computation of ventilation currents as derived from Laws I. and II. Unit head or aeromotive force (·015 inch water pressure) maintains unit flow (1 cubic ft. per sec.) through unit resistance (a circular aperture 6 in. in diameter in a thin plate).

The head required to maintain a flow through a given resistance is proportional to the square of the flow, and the resistance is inversely proportional to the square of the area of the aperture, i.e. inversely proportional to the fourth power of the linear dimensions for thin plate apertures of similar shape.

These complicated algebraical laws may seem to be altogether out of place in dealing with an everyday subject like ventilation.

* Stevenson and Murphy's *Hygiene*, vol. I. p. 58.

They are however unfortunately the laws to which ordinary
ventilation is subject, and the complexity which they represent
is the complexity of real life. It must be allowed for if the
practice of ventilation is to be fully understood. Neglect of
these laws of numerical relation accounts in a great measure for
the failure of ventilation appliances. To take a simple application; if all the apertures for ventilation are doubled in linear
dimension, other things remaining the same, the flow is increased
fourfold, but to increase the flow fourfold, the aperture remaining
the same, the head must be increased sixteenfold.

Law III.   Pneumatic resistances in Series.

We may now proceed to deduce consequences from the
application of Law II. to more complicated cases, and in the first
place to consider the simple circuit as made up of separate
parts, the inlet and the outlet, each having resistance, with an
intermediate ventilated space having no resistance. We will
consider how the work done in carrying air through the circuit
is distributed.

For this purpose we have simply to remember that the work
done in the whole circuit is the sum of the items of work done
in the separate parts; wherever there is resistance work is
done, where there is no resistance no work is done, and the
sum total of work is additive. Neglecting the work done in
getting the air across the ventilated space because there is no
appreciable resistance, we may divide the total head H into two
parts, viz. h_1, that spent in carrying air through the inlet, and
h_2, that spent in carrying it through the outlet; then, since the
total work is additive,

$$H = h_1 + h_2.$$

Next let R_1 and R_2 be the resistances of the inlet and outlet
respectively, R the equivalent resistance of the whole circuit,
and V the flow through the whole circuit.

We have for the flow through R_1 and R_2

$$h_1 = R_1 V^2, \quad h_2 = R_2 V^2,$$

and through the whole circuit,

$$H = R V^2,$$

but since $\qquad H = h_1 + h_2,$

we get $\qquad R_1 V^2 + R_2 V^2 = R V^2,$

or $\qquad R_1 + R_2 = R.$

In other words we obtain, just as with electrical resistances,

LAW III. The equivalent pneumatic resistance of a circuit is the sum of the pneumatic resistances of the several parts arranged in series.

The same law may obviously be extended to apply to any number of ducts in series. An inlet and outlet correspond to wires for the passage of electricity, the ventilated space to a metal block connecting them.

It should be noticed here that the head or aeromotive force for ventilation is produced, generally speaking, in some single part of the circuit, by a fire at the foot of a chimney shaft, a fan in an inlet or outlet, or by wind blowing upon or across an opening. In those circuits which contain several pneumatic resistances in series, the partial head for the flow through one of the resistances, not provided with its own aeromotive force, has to be transmitted, just as the power from a battery in one part of an electric circuit is transmitted along the wire in order to cause the current to overcome a distant resistance. And further, just as with a battery, the transmission of the power may only account for a part of the expenditure of the energy of the battery, the rest being spent in local action, so with pneumatic force, transmission of energy may be very uneconomical in consequence of the local action or waste of energy in the pneumatic agent. This is very conspicuous in the case of low pressure fans, which become very uneconomical if the resistance to be overcome is high. We will return to this point later on.

Law IV. Resistances in parallel, or multiple arc.

Head, or aeromotive force, may be expressed as pressure difference, and as in a ventilated space any temporary pressure difference between two parts would be immediately compensated by flow of air, we may assume that, when there are two

channels "in parallel" connecting two spaces, the head or aeromotive force is the same for both, provided the channels
themselves do not give rise to local aeromotive forces. Hence
H is the same for the two "parallel" channels, and if R_1, R_2
are the resistances of the two channels, we have

$$H/R_1 = V_1^2 \text{ and } H/R_2 = V_2^2.$$

We will now substitute for the resistances R_1, R_2 the areas
α_1, α_2 of the thin plate orifices to which they are equivalent, in
accordance with the equations $R_1 = 1/(27\alpha_1^2)$ and $R_2 = 1/(27\alpha_2^2)$.
We thus get

$$27\alpha_1^2 H = V_1^2 \text{ and } 27\alpha_2^2 H = V_2^2.$$

Then, if R is the equivalent resistance of the two ducts
combined and α its equivalent thin plate orifice,

$$27\alpha^2 H = V^2.$$

But the flow of air in the two ducts combined is the sum of
the separate flow through each, thus

$$V = V_1 + V_2,$$

whence we get

$$\alpha = \alpha_1 + \alpha_2.$$

The same reasoning applies to any number of ducts in
parallel and thus we deduce:—

LAW IV. The combined effect of channels in parallel
arc is the same as that of a single channel whose equivalent
thin plate orifice is equal in area to the sum of the areas
of the equivalent thin plate orifices for the separate channels.

Indirect verification of the laws.

From these four laws, of which the last two are the
necessary consequence of the two former, just as Kirchhoff's
laws of complex electric circuits are the necessary consequence
of Ohm's law and the law of electric quantity, the numerical
relations of complex pneumatic circuits can be deduced. I
propose now to give by way of verification of the laws some
examples of the comparison of inferences from them with actual
experiment in the case of ventilation circuits of various kinds.

Measurement of the effective or equivalent area of a chimney.

The first example is derived from some experiments upon the flow of air in a chimney flue of the Cavendish Laboratory, due to the aeromotive force arising from heated air in the flue. The chimney had at its base a hearth but no fire-grate; heat was supplied by burning gas in the chimney. The opening in front was provided with a close-fitting wooden door, the air was admitted through a series of 3 inch circular perforations, any of which could be stopped when required by means of cardboard covers. The flow of air through any one of the openings could be determined by the readings of a Biram air meter applied at the aperture.

The object of the experiments was to determine the resistance of the chimney by measurements of the effect produced upon the flow by intercalating known resistances in the circuit. The process is exactly analogous to the computation of the resistance of a battery from measurements of the current passing through a galvanometer. Referring for a moment to that process, with the usual notation, we have

$$\frac{E}{B+R} = C.$$

R is the resistance which is made to vary, and C is the corresponding current. Writing the equation in the form

$$B + R = \frac{E}{C},$$

it is evident that the relation between R and $1/C$ is linear, provided E is constant, and therefore a graph representing the relation between R and $1/C$ should be a straight line, and the solution of the equation may be carried out graphically.

The corresponding calculation for the chimney is as follows:—If H is the aeromotive force, A the resistance of the measured apertures at the base, Ω the resistance of the chimney, we get a flow V through the apertures and through the chimney due to H. Assuming that the hearth space is sufficiently large for it to be regarded as a "ventilated space" communicating on the one side through the apertures with the room and on the other through the chimney with the open air, then by Law III. the total resistance of the circuit is $A + \Omega$, and

$$H = (A + \Omega) V^2,$$

or $$A + \Omega = \frac{H}{V^2}.$$

There is thus a linear relation between A and $1/V^2$. A can be computed by Law IV. from the areas of the apertures, regarded as thin plate apertures. V is computed from the areas and the measurements with the air meter, and thus the determination of Ω by a graphic method can be followed if H be constant. And on the other hand the existence of a linear relation between the measurements of A and H/V^2 would be evidence of the truth of the law of flow for the particular circuit.

In the electrical problem it is safe to assume E constant, but in the corresponding pneumatic case, where the aeromotive force depends upon difference of temperature produced by a constant supply of heat to the flowing air, we cannot make a similar assumption, and arrangements were accordingly made for determining the variation of H by measuring the difference of temperature $(T - t)$ between the air in the flue and that outside the window. We then assume that $H = k(T - t)$ where k is a constant. We thus get, if the laws be true, a linear relation between $\frac{T - t}{V^2}$ and A, from which Ω can be determined graphically.

A series of experiments to illustrate this method of determining the resistance of a chimney was carried out for me by Mr J. W. Madeley of Emmanuel College in the Long Vacation of 1891, and I quote the following paragraphs from his account of the work.

Experiments on the flow of air up chimneys.

For this purpose an ordinary chimney was employed, and the head was produced by burning gas: the front was covered by a wooden screen, rendered as air-tight as possible, in this were a number (6) of circular apertures all of which could be closed, separately or together.

The head was determined by finding the difference in temperature of the air in the chimney and that outside, the difference in temperature being determined by means of a differential platinum thermometer. One coil was suspended half-way up the chimney and the other outside the window.

The windows of the room were kept wide open in order that none of the head should be used up in drawing the air through the chinks in the windows and doors.

The general arrangement is shown in the accompanying diagram, Figure 1: o represents (diagrammatically) the aperture

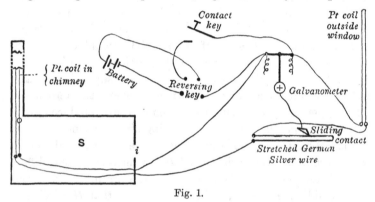

Fig. 1.

of the chimney, S the space at the foot of the chimney inside the screen, and i the adjustable aperture in the screen.

The flow (V) was determined from the area of the apertures and the velocity of flow: the area of the apertures being measured directly and the velocity measured in feet per second by means of the air-meter.

From the data thus determined, i.e. the resistance of i, denoted by A, the difference in temperature* ($T-t$) of the air in the chimney and that outside, and the flow (V) of air up the chimney, we can calculate the resistance of the chimney Ω and thence deduce the area of the equivalent thin plate aperture.

* In what follows the temperature of the air in the chimney has been taken as uniform, and as given by the temperature of the platinum wire supported at the top of a fishing-rod about 10 feet long.

Where $\mathbf{H}$ is the total head, $\mathbf{H}_1$ the head required to force the air through i, and $\mathbf{H}_2$ the head used up at o.

$$\mathbf{H} = \mathbf{H}_1 + \mathbf{H}_2$$
$$= (A + \Omega) \, V^2 \, ;$$
$$\therefore \frac{\mathbf{H}}{V^2} = A + \Omega.$$

Also $\qquad \mathbf{H} = k \, (T - t),$

where k is a constant,

$$\therefore \frac{k \, (T - t)}{V^2} = A + \Omega,$$

which may be written in the form $ky = x + \Omega$ (representing a straight line), but

$$k = \frac{dx}{dy},$$
$$\therefore y \frac{dx}{dy} = x + \Omega,$$
$$\therefore \Omega = y \frac{dx}{dy} - x.$$

Hence if we draw a curve PST, Figure 2, the ordinates of which represent $\dfrac{T - t}{V^2}$, and the abscissae represent the different values of A, we shall get a straight line, and if this be produced to meet the axis of x in T, OT will represent the resistance of the chimney.

We shall require a correction for the variation of the density of the air on account of the change of temperature due to the heating.

The method of introducing the correction is shown as follows :

As before $\quad \mathbf{H} = \mathbf{H}_1 + \mathbf{H}_2$

$$= A \, V_1^2 + \Omega \, V_2^2$$
$$= \left\{ A + \left(\frac{\rho}{\rho'} \right)^2 \Omega \right\} \, V_1^2 \quad \text{(for } V_1 \rho_1 = V_2 \rho_2 \text{)}.$$
$$\therefore \frac{\mathbf{H}}{V_1^2} \left(\frac{\rho'}{\rho} \right)^2 = A \left(\frac{\rho'}{\rho} \right)^2 + \Omega.$$

But $$\frac{\rho'}{\rho} = \frac{1 + at}{1 + aT},$$

where a is the coefficient of thermal expansion of air,

$$\therefore \frac{H}{V^2} = A + \frac{\Omega}{1 - 2a\,(T - t)}, \text{ approximately.}$$

Hence, as before, OT represents Ω, or

$$\Omega = \{1 - 2a\,(T - t)\}\,OT.$$

Hence, as before stated, in order to correct the computed resistance for variations in density due to differences in the temperature of the air, all we have to do is to multiply the result by $\{1 - 2a\,(T - t)\}$.

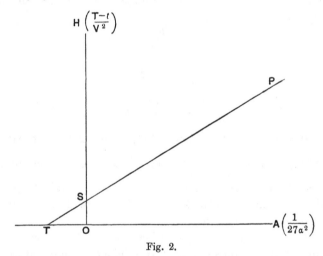

Fig. 2.

The results of two of the sets of experiments are represented on a diagram, Figure 3. In the first set the temperature of the air in the chimney, and consequently the head, was allowed to vary with the flow and the results were accordingly subject to a correction of considerable magnitude. In the last two series the gas was adjusted so as to keep the temperature constant throughout the measurements, and it is from these two that the results represented in the diagram were obtained. The resistance A, of the aperture in the screen in front of the chimney corresponding with the opening of a number of holes in succession from one up to six, computed by the formula $A = 1/(27a^2)$, is shown along the line of abscissae, the ordinates are the corresponding values of $(T - t)/V^2$, i.e. the head divided by the square

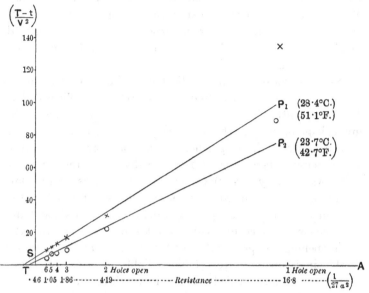

Fig. 3.   X Experiments of the first series.   O Experiments of the second series.

of the flow. P_1ST and P_2ST are drawn to represent the straight lines deduced from the observations, one for a difference of temperature of $28\cdot4°$ C. and the other for $23\cdot7°$ C. The agreement between the observations and these lines, or others that might be drawn instead, is the test of the appropriateness of the law in this case.

While the results cannot be put forward as specimens of experimental accuracy, they are very suggestive illustrations of the analogy between the pneumatic problem of finding the resistance of a chimney from observations of the flow and the electrical problem of finding the resistance of a battery from observations of the current which it produces. The straight lines drawn in the diagram are obviously rather arbitrary but they are not altogether astray. They do not nearly agree with the observation of flow through a single hole, but considering the circumstances the agreement is as good or better than one might have expected. In the first place the chimney opens to the free air and any wind over the top would produce a casual aeromotive force supplementing that allowed for in the measurements, so that specially calm days had to be selected for making the experiments, and even careful selection could not eliminate the risk of interference arising from the wind.

Next the measurement of the flow and area for the single hole is the one most liable to error. It cannot be affirmed that the shutter fitted the chimney opening without leakage, and any leakage or error in the measurement would affect the point corresponding to the single hole more than any other. Since the measurements were made I have realised that the air meter which was employed in the experiments to measure the flow should have a different constant according as it is used in the open air or in a tubular opening which it nearly fits*, and any allowance on this account will affect the computation from the single opening but practically not the others. Hence the values for the single opening are peculiarly liable to error and ought perhaps to be disregarded. If we leave these out of account the other observations group themselves fairly well in straight lines.

It is however clear that the dimensions of the chimney are so considerable that the method of determining its resistance by computation from the flow is not shown at its best. We ought for accuracy to have had larger openings and have worked with greater minuteness, but we cannot choose in these cases and the agreement of the two sets of observations is very reassuring. The analogy with electrical methods is thus well made out. When allowance is made for the variation of density of the air in the flow the resistance of the chimney computed numerically from the observations works out at ·46 unit, about half of that of six holes, as may be inferred from the diagram. This corresponds to a thin plate orifice of about 40 square inches and represents the practical determination of the thin plate equivalent of a chimney 14 × 9 inches in area, about

* See Report on Cowls and Terminals. *Journal of the Sanitary Institute*, vol. xxii. p. 232, 1901.

20 feet high, with a double bend in its length, and a chimney-pot mouthpiece.

As the resistance is, roughly speaking, half a unit, it follows that unit aeromotive force acting in the chimney would produce flow of $\sqrt{2}$ cubic feet per second.

Experiments of the Cowl Committee of the Sanitary Institute.

The second illustration of the indirect verification of the law of flow of air is taken from the Report on the work of the Committee appointed by the Sanitary Institute of Great Britain to test ventilating exhaust cowls, which is contained in volume XXII. of the *Journal* of the Institute.

In the course of their investigation the Committee made a large number of experiments upon the upcast flow of air in a vertical shaft 3 inches in diameter produced by wind passing the freely exposed end of the shaft. At the same time the velocity of the wind was measured, and the velocity of the upcast compared with that of the wind. The general arrangement is shown in Figure 4 which is taken from the Report. The lower part of the pipe was surrounded by loose fitting boards forming a hut which protected the measuring apparatus and the observer. The lower end of the shaft was covered by a box with a silk gauze screen and at the foot of the pipe, above the box, was an air meter to measure the flow. Thus the resistance to be overcome by the flow was (1) that due to the gauze and the entry to the vertical shaft, (2) that due to the passage of the air past the meter, and (3) that due to the opening at the top of the pipe. The results obtained are shown in Figure 5. The particular curve to which I wish to call attention is the "velocity of upcast in centre pipe." The abscissae represent the velocity of the wind as determined by a Dines pressure tube.

The pneumatic equation giving the expenditure of work to maintain the upcast flow must be a complicated one if all considerations are to be taken into account. For the head we have the effect of blowing across the top of the pipe which may be taken as proportional to the square of the wind velocity, but

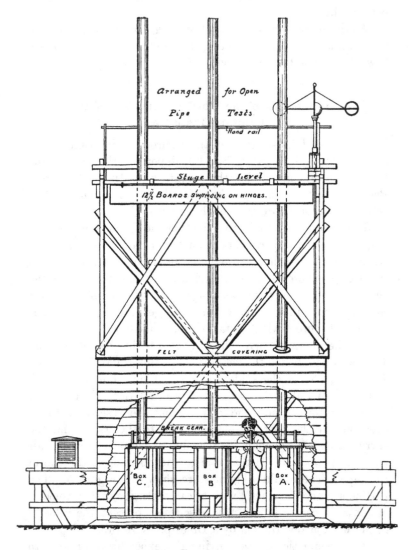

ELEVATION

Fig. 4. General view of arrangements by the Cowl Committee of the Sanitary
Institute for experiments upon the upcast flow of air caused by wind.

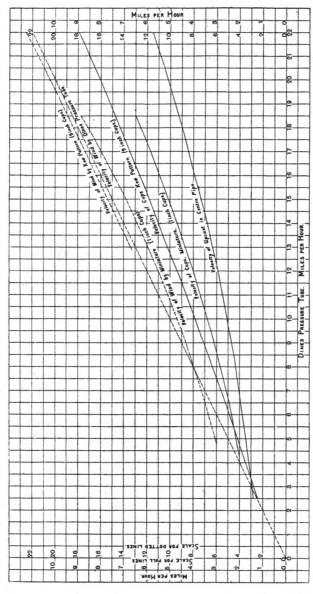

Fig. 5. Robinson Cup and Dines Pressure Tube anemometer compared with upcast in centre pipe.

there may be other sources of head, such as a difference of temperature of the air inside and outside the shaft; thus the total aeromotive force would be represented by H_w (depending on the wind velocity v, and equal to kv^2) + H_x (depending on temperature or other undefined causes). The work would be spent partly in overcoming the pneumatic resistance proportional to V^2, the square of the upcast flow, and partly, since the pipe is long and comparatively narrow, in overcoming frictional resistance proportional to the first power of the flow.

Thus the work spent may be represented by

$$RV^2 + rV \quad \text{or} \quad RV^2\left(1 + \frac{\mu}{V}\right).$$

Hence the full equation would be

$$kv^2 + H_x = RV^2\left(1 + \frac{\mu}{V}\right).$$

This equation would be reduced to the simpler one if we could neglect H_x the head due to temperature differences, and μ the term depending upon frictional resistance. We should then obtain the simple relation

$$kv^2 = RV^2,$$

or V proportional to v, the upcast simply proportional to the wind.

On examining the curve it will be seen that this equation does represent the relation with quite sufficient accuracy for wind velocities beyond 6 miles per hour; below that velocity the upcast is greater than the continuation of the linear relation would indicate. Now the frictional resistance would gradually bring the point representing the upcast below that representing the linear relation at high velocities, whereas the undefined head H_x due to temperature would take it above. It will be noticed that the curve if continued would cut the vertical axis through the origin at a velocity of about one mile per hour, a finite velocity with no wind. Consequently the effect of the undefined head predominates. Without going into the question numerically, it will be sufficient to say that the course of the curve—representing the upcast in relation to the wind—is consistent with the supposition that the term due to friction in the pipe is negligible, that the pneumatic law $H/V^2 = R$ holds, and that the head H includes, besides that due to wind, a portion H_x, depending probably upon a difference of temperature between the pipe and the outside air, sufficient to give an upcast flow of about one mile per hour when there was no wind.

So much natural ventilation depends upon the combined effect of wind and temperature that this illustration of the combination of the two effects may be taken as typical of a large number of cases occurring in practice.

The Pneumatic Analogue of the Wheatstone Bridge.

The third illustration of the indirect verification of the pneumatic laws of flow is susceptible of much greater accuracy than either of the two already mentioned, and morever, it makes a further step in the development of the electrical analogy because it introduces a null method into pneumatic measurement.

We have seen that apertures in pneumatics correspond with conducting wires in galvanic electricity; large boxes, which are put into communication by the apertures, correspond with the brass blocks and plugs, or binding screws, which join wires together. Thin plate apertures are those which give the most constant ratio H/V^2; and consequently large boxes with inter-communication by means of thin plate apertures give the best material with which to construct the pneumatic analogues of electrical apparatus.

Four boxes forming a pneumatic quadrilateral, with thin apertures between consecutive pairs, give a quadrilateral of pneumatic resistances. A tube between two opposite boxes containing a very delicate indicator of air-flow corresponds with the bridge wire and galvanometer of the Wheatstone quadrilateral; as the method is to be a null method, that is, as things are going to be so adjusted that no air flows along the tube, there is no disadvantage in using a tube conductor instead of a thin plate aperture.

To complete the equipment in the analogy of the Wheatstone bridge we require an aeromotive force between the pair

of opposite boxes not connected by the tube, to correspond with
the battery.   There would not be much difficulty about providing
this; but in practice it is easier to use the free air instead of the
fourth box, and then two of the thin plate orifices will be open
to the air, and the aeromotive force can be introduced between
the cross-box and the free air.   This is a convenient arrangement in two ways; first one can get at the two apertures to
alter or adjust them, and secondly one can use a vertical metal
tube with a gas jet in it to give the necessary aeromotive
force.

I constructed an apparatus of this kind in 1890 and it is
described and figured in the *Proceedings of the Royal Society*,
vol. XLVII. p. 465.   The diagram is reproduced in Figure 6.

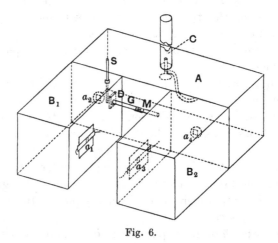

Fig. 6.

The theory is quite on the lines of that of the Wheatstone
bridge, except that, in the pneumatic analogue, the square of the
flow is to be taken instead of the first power of the current.
It is briefly as follows :

Let H be the aeromotive force due to the heated air in the
chimney C.   The flow which supplies it comes partly through
the apertures a_1, a_2, and partly through a_3, a_4.   If R_1, R_2, R_3,
R_4, are the several resistances, h_1, h_2, h_3, h_4, the "partial"
aeromotive forces, V_1 the flow through the one pair of apertures,

V_2 that through the other pair, supposing there is no flow
along the tube DM between the opposite boxes we have

$$V_1{}^2 = \frac{h_1}{R_1} = \frac{h_2}{R_2}, \quad V_2{}^2 = \frac{h_3}{R_3} = \frac{h_4}{R_4}.$$

By law 3, $h_1 + h_2 = h_3 + h_4.$

And since there is no flow in the connecting tube

$$h_1 = h_2 \text{ and } h_3 = h_4,$$

whence $$\frac{R_1}{R_2} = \frac{R_3}{R_4},$$

a relation which corresponds with that of the Wheatstone
quadrilateral.

The apparatus described in the paper referred to was
modified, so as to give the means of adjusting one of the
openings with great accuracy. Many measurements were thus
made of the pneumatic resistance equivalent to thin plate
orifices of various shapes, and their combinations, as well as
of tubes of various diameters and lengths.

The complete account of these measurements, which were
carried out by Mr R. S. Cole of Emmanuel College, has not yet
been published. The precision which was obtained may fairly
be called remarkable. It was proved thereby that with thin
plate orifices the bridge conditions are independent of the
aeromotive force, as in the electrical analogy; that the resistance
of a thin plate aperture depends on its size and not on its shape;
that combinations of thin plate orifices are equivalent to a
single orifice of the same total area although when the number
of orifices is very large as in the case of perforated zinc the
collection of orifices is a little more effective for admitting air
than a single orifice of the same aggregate area.

The best illustration of the accuracy of which such measurements are capable is shown by the results for various lengths of
tube, and the effects of a flange or coned end. These results are
represented in a diagram, Figure 7, and will be understood
without further description. It is evident that after a certain
length of tube is reached the flow no longer follows the simple
rule of resistance proportional to H/V^2; the frictional term
begins to produce effect. So regular is the effect that the

coefficient of friction can be determined from the measurements in a manner which is reasónably satisfactory considering the incidental difficulties.

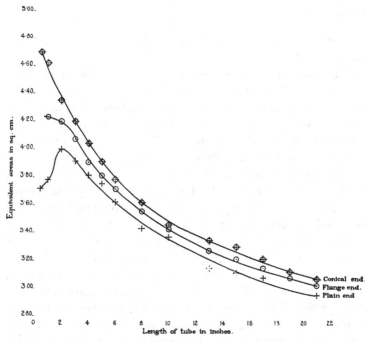

Fig. 7. Diagram of equivalent areas of different lengths of tube.

These experiments furnish the most striking confirmation of the accuracy of the pneumatic laws.

The Pneumatic Analogue of the Potentiometer.

The fourth illustration of the pneumatic laws is also an example of a null method and refers to an apparatus which represents the pneumatic equivalent of the potentiometer. It

was exhibited at the meeting of the British Association at
Bristol in 1898 and is represented in Figure 8. In this case

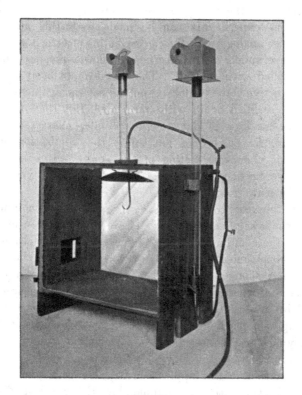

Fig. 8.

two aeromotive forces, produced by small gas jets in vertical
glass tubes, act upon the same box which has a sliding shutter,
a third communication with the outside of the box. In the
model the box has glass sides which do not show in the figure.
The flow detectors in this case are plates of mica framed in
square tubes and very lightly balanced on knife edges. They
are indicated by the square boxes shown at the tops of the
tubes in the figure.

In strict accordance with the electrical analogy, as the
sliding shutter is gradually closed the flow up the tube which

contains the smaller aeromotive force becomes less and less until it is altogether arrested and finally reversed. The air goes down the tube when the ratio of the aeromotive forces is over-passed by the adjustment of resistances.

Various experiments can be made with this apparatus. The precise analogy with the corresponding electrical apparatus is, of course, rudely disturbed when the flow is reversed and the tube is filled with cold air. The reversal obviously removes the aeromotive force, which cannot be set up again until the tube can be filled again with warm air. But that departure from the analogy is due to the peculiar character of the aeromotive force; it does not affect the application of the analogy to the consideration of the compensating effects of aeromotive forces such as those of two chimney flues acting upon the same room.

It is thus easy to adduce many illustrations of the application of the laws of ventilation which we have been considering; some of those adduced show the conditions under which the law of proportionality between aeromotive force and the square of flow is no longer applicable. From the nature of the case none of the experiments which depend upon air measurements can claim anything like the accuracy of Ohm's law; but they are amply sufficient to give a general indication of what actually goes on in the processes of ventilation and to explain many important facts in connexion therewith.

Illustrations of the application of these principles in connexion with ventilation are innumerable. Every pair of adjacent rooms with a fire in each and a communication between them, may at any time furnish an example similar to the analogue of the potentiometer; for if the resistances are suitably adjusted, the aeromotive force of the more powerful chimney will reverse the flow in the less powerful. In the article on "Ventilation and Warming" in Stevenson and Murphy's *Hygiene* I have given a number of calculations based upon these principles. Briefly every house or building considered from the point of view of its ventilation is an extraordinarily elaborate combination of aeromotive forces, due to wind acting in various ways, temperatures in flues, and sometimes blowing fans in addition, with resistances of chimneys, ducts, chinks, gratings, and possibly also of large areas of porous walls. On the electrical analogy the whole presents a complication of aeromotive forces and resistances, generally of a varying character, the interrelation of which no electrical engineer would care to specify in figures. It is little wonder therefore that the computation of the figures for ventilation under similar conditions has never yet been attempted.

Yet the analysis of the laws of flow of air currents which has been presented here is not intended to indicate that the complexity is so great that consideration of the subject from this point of view is valueless. The consideration will be a guide, in the first place, to the advantages of simplification of systems, and to the methods of achieving it; and secondly, though precise numerical computation may be too difficult to be attempted, the mental pictures of the ways of air currents in relation to the known conditions will give at least a qualitative guide to what may be expected in actual practice.

I was once asked to report upon a scheme of ventilation for the out-patient department of a large Hospital. There were something like 300 rooms of one kind or other on four floors, and a variety of appliances for warming and ventilating them. I proceeded to draw the system of circuits on the basis of the electrical analogy, and, as may be imagined, the complexity of wires and batteries required to represent the air circulations was hopelessly entangled. Still it faithfully represented the complexity of the actual case and it was comparatively easy to estimate the importance of this or that disturbing element, which might easily have been overlooked if I had attempted to simplify the problem by neglecting all those elements of it that come in, whether we desire them or not, as disturbances of the adopted system.

The Interference of air-flows from Pipes delivering into a Common Conduit.

Let me supplement what has been said already by an additional example of the application of the electrical analogy, in which, although perfectly arbitrary magnitudes of the quantities are used (for want of accurate knowledge of such subjects) the method of treatment throws a good deal of light on the problem.

A few weeks ago my opinion was asked as to the probable effectiveness of a system of ventilation, for a series of fume-closets in a chemical laboratory, by carrying vertical pipes from the fume-closets to a horizontal pipe which passed across the series. Aeromotive force was to be supplied to the horizontal pipe by a fan in a chamber with which the pipe communicated.

Figure 9 represents the pneumatic distribution. A is the

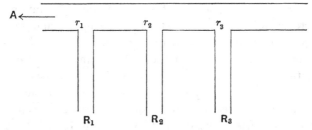

Fig. 9. Vertical ducts delivering into a horizontal duct.

aeromotive force due to the fan; R_1, R_2, R_3 the resistances of the orifices leading from the fume-chambers. The special point about this problem is that when the air which has passed the resistance R_1 reaches the larger pipe an expenditure of energy will be required to make it turn the corner towards the fan. That this is so will be easily understood if we consider what would happen if the pipe leading from R_1 were continued into the wider tube and bent so as to face A, or away from it. Clearly in the latter case the issuing air would face the flow in the pipe coming from R_2 and R_3 and great resistance would ensue, much more than in the former case, i.e. if the bent pipe delivered the air towards A. Thus the flow of air from R_1 into the pipe causes resistance in the main circuit; for this there is no counterpart in electricity. Therefore in drawing a diagram for the electrical analogy, as in Figure 10, we must introduce on this account resistances r_1, r_2, r_3 in the main circuit due to the currents from the vertical tubes. These resistances clearly depend upon the size of the horizontal pipe, being less the larger the pipe. If the pipe were very large they could be

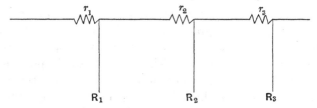

Fig. 10. Electrical analogy of Fig. 9.

neglected, or would appear simply as additions to the resistance of the vertical pipes. If the horizontal pipe were narrow the resistances r_1, r_2, r_3 produced in the main circuit would be very considerable.

Suppose further that these resistances can be regarded as occurring just beyond the orifice of the pipe R_1, so that in going round the circuit $R_1 r_2 R_2 R_1$ the resistance r_1 will just be missed.

Applying the pneumatic analogue of Kirchhoff's law to this circuit and neglecting every aeromotive force in these parts of the circuit, we get, if V_1 is the flow through R_1 and V that in the main circuit,

$$R_1 V_1^2 - r_2 (V - V_1)^2 - R_2 V_2^2 = 0.$$

We know nothing definite about the magnitudes of R_1, r_2, or R_2; for simplicity suppose them all equal, then

$$V_1^2 - (V - V_1)^2 = V_2^2,$$
$$\therefore \ V(2V_1 - V) = V_2^2.$$

Thus $2V_1 - V$ cannot be negative; and therefore $2V_1$ cannot be less than V, or V_1 cannot be less than $\frac{1}{2}V$.

Hence under the conditions assumed, however many fume-closets there were connected up, at least one-half of the whole air supply to the fan A would go through the nearest duct; at least one-half of the remainder through the second, and hence the distribution of air flow between the series of ducts must be more unfavourable than the distribution $\frac{1}{2}, \frac{1}{4}, \frac{1}{8}, \frac{1}{16}, \frac{1}{32} \ldots$ for the closets in succession.

If there were two closets only, we should get $V_1 = \sqrt{2}\,V_2$ and thus five-eighths of the flow would go through the first and three-eighths through the second.

It is easy to say that the assumptions upon which this calculation is based are gratuitous. The result is a sufficiently accurate representation of what actually happens when a number of ventilation pipes deliver into a single pipe, in the manner indicated, to make one curious to know what the actual numbers are which ought to be substituted for the hypothetical ones; and I know of no other suggestion which gives so effective an explanation of the interference of one pipe with the action of another.

There can be little doubt that the experimental investigation of the flow of air, guided by the electrical analogy, would help to clear our ideas about many important points in connexion with ventilation, but the problems arising in relation to the distribution of air in the ventilated space will claim our attention next.

II.

PHYSICAL PRINCIPLES APPLICABLE TO THE VENTILATED SPACE.

Energy and Momentum.

I WILL first briefly call your attention to certain considerations with respect to the energy and momentum of the air which enters and leaves a ventilated space; and, for the purpose of fixing ideas, let me refer to a diagram, Figure 11, which represents

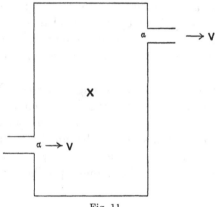

Fig. 11.

in a generalised way the openings for supplying and removing air from the space X. I call the area of the orifice, by which air enters or leaves the space, α; the flow, inwards or outwards as the case may be, is represented by V; the density of the air by ρ. The use of identical symbols for the air at two different places does not necessarily mean that the quantities have the same numerical values at the two places. If the reasoning were to be applied to a real case, with specified dimensions, proper numerical values would have to be supplied for the two places.

As to energy :—The velocity of the incoming air is V/α, the amount that comes in in a second is $V\rho$; hence the energy represented by the jet is $\frac{1}{2}\rho V^3/\alpha$ per second.

That energy is dissipated in the ventilated space. The corresponding energy of emergence, $\frac{1}{2}\rho V^3/\alpha$ per second, has to be obtained separately by an additional draft on the supply of power. Both are wasted if, and in so far as, the flow might be produced with less expenditure of energy, because velocity in the ventilated space is of no advantage. I call attention to two points : (1) *doubling V* keeping α constant means taking *eight times the power*, and (2) *doubling the aperture* and keeping *V* constant reduces the power required by *one-half*.

It follows that large apertures are of great importance for economical ventilation. Thus, if it is necessary to increase the supply of air to a ventilated space, the proper plan is to increase the openings; to leave the openings unchanged and endeavour to produce the requisite flow by increasing the motive power is, under ordinary circumstances, an extravagant expenditure of energy.

The momentum of a jet is not long preserved. The object of ventilation may be described as the production of change of air without perceptible momentum. For momentum, the law as regards ventilated spaces should be an instruction to keep it within the smallest possible limits by providing for the distribution of air over the largest possible area before it enters.

Convection.

The dominant physical law in the ventilated space is the law of convection. It is at once the condition of success and the cause of most failures. Without convection ventilation would be impossible; in consequence of convection nearly all schemes of ventilation fail.

The law of convection is the law according to which warmed

air rises and cooled air sinks in the surrounding air. Its applications are truly ubiquitous. Every surface, e.g. a warm wall, or a person warmer than the air in the immediate neighbourhood, causes an upward current; every surface colder than the air in contact with it causes a downward current. The convection currents are interfered with by momentum currents; but the momentum currents only interfere seriously with the horizontal distribution in the middle regions of a room. The vertical distribution of air currents in a ventilated space, and their horizontal distribution at floor and ceiling are to be ascribed generally to convection. Ventilation would be much easier if warmed air or cooled air could be carried along at any height required; but the law of convection is inexorable, warmed air naturally finds the ceiling, cooled air the floor.

Apparatus.

I wish to call attention to some apparatus designed to identify the flow of air in the ventilated space and thereby to illustrate how dominant the law of convection is. It consists of some very sensitive wind vanes, balanced on needle points which are supported by small glass caps, and wheels consisting of inclined plates of mica mounted in cork round glass caps, which are also balanced upon a needle point and show the motion of the air by the rotation of the wheel. The apparatus was designed for the investigation of the ventilation of Poor Law Schools for the Local Government Board in 1897; an extensible stand to carry these and other apparatus, such as thermometers, at any height in a room was designed at the same time. The stand carrying the wind vane and mica wheel is shown in Figure 12. The vane is shown on the left of the stem, the wheel on the right; the thermometer is held in a perforation through the stem and wedged in by a doubled elastic band. The stem of the stand can be extended by additional lengths joined to the base length by a brass socket. The little wind vanes show the horizontal movement of air, the balanced mica wheels the vertical movement.

With these simple contrivances the circulation of air is

easily identified; they are extraordinarily sensitive; it is easy, for example, to show by the mica wheel the ascending current due to the warmth of the hand. Investigations with the apparatus

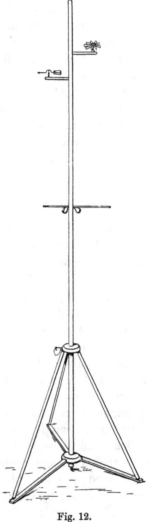

Fig. 12.

have led to some results which are remarkable at first sight, but cease to be so if we allow for the *inexorable character of the law of convection.* For example in a certain room, which I know very well, there is always in cold weather a cold floor current on the side of the fire away from the window—I say away from the window because the draught on the side of the fire near the window needs no elaborate explanation. I always supposed, and my friends concurred in the opinion, that any cold air coming from the window would be compelled by *force majeure* to go up the chimney; but a closer investigation showed a steady up-current on the wall *beyond* the fire-place due to the warmth supplied through the brickwork by a neighbour's chimney; and of course the cold supply from the window passing across the floor had to furnish not only the chimney of the room but the convection current on the wall beyond the fire-place. One might have supposed that the fire would have stopped the cold air on its way to the wall, but a little reflection shows that that arrangement would not satisfy the warm place on the wall.

The fact is that in nearly all ventilation schemes there is a sort of sub-conscious idea that if you give nature sufficient occupation in dealing with the main principle of the scheme she will not have time or ability to keep these little disturbing causes in working order. No idea could be more fallacious. It is no trouble to her to take care that a large cold window produces a chilling down-draught, and that a little patch of sunlit wall causes an up-draught: her attention is not distracted by the fact that a four horse-power engine is exerting itself to the utmost to keep the ventilation of the building in its normal channel. She provides without difficulty a convection current wherever air is locally warmed.

Momentum currents are sometimes directed in opposition to the forces of convection, and for a short distance the momentum current is operative; but its effect soon ceases (see Figure 13) and the convection current generally speaking rapidly asserts itself. It not unfrequently happens that the momentum current consists of air at a temperature conspicuously different from that of the surrounding air, and in consequence a convection current is soon set up. The draught thereby produced is often attributed to the momentum of the entering current, whereas if the temperature had been adjusted so that no convection current was produced, or, if produced, was directed upward instead of downward, the momentum would have escaped without being accused of causing the draught which is rightly attributable to convection.

The ubiquity and activity of convection currents is most easily illustrated by the rapid distribution of the smell of tobacco smoke, or some easily identified vapour, about a fully occupied room. The experiment may be tried with a fumigating pastille, which, if burned in one part of the room, rapidly makes itself perceptible throughout the room. Formerly it was usual to attribute this effect to the process of diffusion alone; but diffusion is an incomparably slower process than convection in an occupied room. It would of course be difficult to attribute the effect to convection if no sources of heat were present to produce convection currents; but when we have shown, as by the experiment with the mica wheel, that the warmth of a hand produces a vigorous convection current, and infer therefrom that every person is an active convection agent, and when we add the effects of fires, warm walls and hot water pipes, it is altogether unnecessary to invoke the aid of diffusion, on a scale enlarged far beyond its natural capacity, to account for the rapid dissemination of the pastille smoke throughout the room. Diffusion may be safely left to act as a modest assistant of the more powerful convection.

Experimental results.

Some results of the direct observation of air currents in a variety of interesting cases are given in the diagrams which follow, Figures 13, 14, 15, 16. They were originally prepared for the article on Air—"Ventilation and Warming" in Stevenson and Murphy's *Hygiene* and are reproduced from the drawings in that work by permission of Messrs J. and A. Churchill. In all of them the effects of convection are clearly indicated. The first, Figure 13, shows the reversal of the direction of a momentum current by the convection due to the low temperature

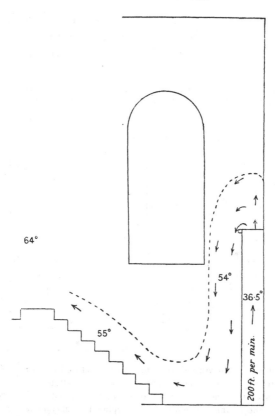

Fig. 13. Diagram showing the flow of air from a Tobin tube in the Biological Lecture Room in Cambridge. The air at a temperature of 36·5° F. is projected vertically upward with a velocity of three feet per second into a room which gives an indication of 64° on a thermometer a long way from the Tobin. The direction of flow of the cold air is reversed about three feet above the opening of the Tobin. The temperature is increased, either by mixture or otherwise, so that 54° is indicated in the descending stream, which represents the continuation of the cold air-flow. But the convection current can be traced from there to the ground, filling a hollow formed by a short stairway, and ultimately flowing over the top of the stairway into the main space of the lecture room. Very little change of temperature is shown in this part of its course. A Tobin tube in the opposite corner of the room, which delivered warm air at a less velocity than the cold Tobin, showed on the other hand a rising current, which continued up to the ceiling, and then spread itself over the ceiling as a thin layer of warm air.

of the entering air. The second, Figure 14, shows the strong convection currents due to the heating apparatus in the Lecture Room of the Cavendish Laboratory. The third, Figure 15,

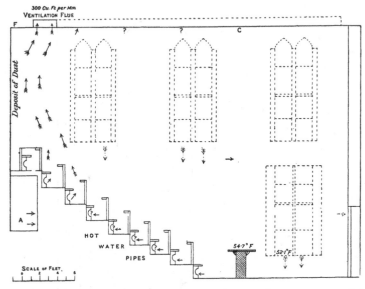

Fig. 14. Diagram showing air currents in the Lecture Room of the Cavendish Laboratory. The strong current shown on the left is a direct convective flow between the hot water pipes under the tiers of seats and the ventilation duct in the ceiling. The air which supplies this current is derived partly from openings at the back, marked *A*, and partly by an inflow of air through openings under the lower seats, which were intended to supply the room with air from the warmed space occupied by the pipes. The warming produces a particularly vigorous current through the openings under the two top rows of seats over the hot pipes. Downward convection currents of cold air are also shown by arrows drawn downward from the window-sills. Hardly any indication of flow into the long ventilation flue, which has openings throughout its lower face, was shown, except where the air rose from the heated pipes.

gives the convection currents on the Laboratory staircase. The fourth, Figure 16, shows the currents in a room heated by an open fire; it is taken from the Report of the Parliamentary Commission on Warming and Ventilation of Dwellings, 1857.

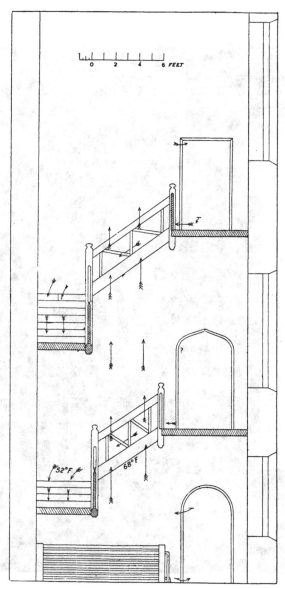

Fig. 15. Diagram showing air currents on the well-staircase of the Cavendish
Laboratory. The main features are a warm upward current from a coil of
pipes at the foot of the staircase, and a cold stream flowing, from the top
story, down the stairs to the basement and thus conducted by the staircase
twice round the rising column of warm air.

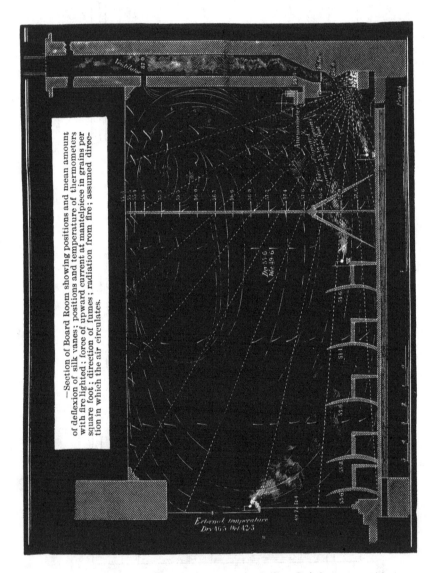

Fig. 16.   Diagram showing the convective circulation of air in a room with an open fire; from the Report of the Parliamentary Commission on Warming and Ventilation of Dwellings, 1857.

Supply of Warm and Cold Air.

I should like to point out how the considerations which I have put before you affect the question of the supply of warm and cold air. Let me point out one of the consequences of the domination of convection.

I have been very frequently asked by those who are distracted with the competition of opposing schemes of ventilation whether I think fresh air ought to be supplied high up or low down in a room. My reply is invariably another question—Are you going to supply it warm or cold? If you supply it warm it will go to the top wherever you put the opening; and on the other hand, if you supply it cold, it will find its way to the floor. The answer to the question depends entirely upon the extent to which you can manage the convection. There is no particular advantage in having the transference from floor to ceiling take place in the room itself so that as regards the final result it does not much matter where you deliver it but it does matter what its temperature is *.

Fresh air and used air.

We have to deal differently with the cases of convection of *fresh air* and convection of *used air*. As regards fresh air we need not trouble about variations of density owing to variations of composition, but with used air the products of combustion in the human body affect the density of the air, by loading it with carbonic anhydride and water vapour. In considering the convection of used air we have to take account of these alterations of composition as well as the temperature. The effect of the carbonic anhydride is to make the air heavier; that of the water vapour and rise of temperature to make it

* In any remarks upon the effect of convection upon the arrangements for ventilation I have assumed that the space to be ventilated is not illuminated by gas. Gas burners give rise to a layer of very hot and impure air at the ceiling which must be removed somehow without bringing it downwards. To pour warmed fresh air into the impure layer at the top of a gas-lighted room would be simply to waste it.

lighter, so that the one may counteract the other two. Whether the expired air will rise or sink depends upon its own temperature, its carbonic acid, its water vapour, and the temperature of the surrounding air. I have made a little calculation about this; I am not sure that the data are correct, but if they are not the calculation could be altered to suit more accurate data. I take the temperature of expired air 98° F., and its whole pressure to be made up approximately as follows:

4 p.c. CO_2
15 p.c. Oxygen
76 p.c. Nitrogen
5 p.c. Water vapour

The result of the calculation is interesting: the expired air will rise if the temperature of the surrounding air is below 81° F.: it will sink if the surrounding air is above 81° F.: it will neither rise nor sink if the surrounding air temperature is 81° F. I will not guarantee the precise accuracy of the figures, but there certainly is a critical temperature beyond which downward convection goes on, and at which the disposal of bad air by convection ceases. At such a temperature a crowd would inevitably be poisoned by its own breath, and a single individual requires a punkah to do mechanically what is done for us in this climate by natural convection currents. In air below this critical temperature each individual acts as his own ventilating agent, through the upward convection currents which he produces. The expired air rises and its place is taken by the surrounding air; the air warmed by contact with the skin or clothes also rises, and thus each person becomes the centre of local heating, and therefore local convection, by which he supplies himself with fresh air, if there is fresh air for him to draw from.

Imagine what would happen at or above the critical temperature to an individual lying in bed. The expired air lies about him forming a layer on his bed until it pours over the edge and falls to the floor.

If you wish to get an idea of what would happen if convection did not exist, you have only to turn your attention to what happens in water the density of which cannot be appreciably affected by the processes going on in the bodies living in it. If a fish were to breathe in water as we breathe in air, i.e. take in water and breathe it out again, there would be no convection because there is no change of density—he would simply form a cloud of used water in front of him which he would breathe over and over again. He adopts another plan, he throws the used water out behind his head; and I venture to suggest that the preference of a fish for gills instead of lungs is due to the fact that he cannot take advantage of convection to get rid of the used material as we do; and I therefore repeat. that, for the purpose of ventilation, convection is the primary condition of success though it is the cause of many failures.

Automatic records of Temperature.

I give next some examples of the results of the convection of used air derived from automatic records of temperature. Unless ventilation is very vigorous the warmth of the respired air is gradually communicated to the whole space, and raises the temperature. If respired air is removed heat is removed with it. Thus the variation of temperature in a building can be used as an index of the working of the ventilation arrangements. The method is really only applicable in the case of a crowded building when the window space is not too large; otherwise the loss of heat through windows spoils its effectiveness. The best way of obtaining the temperature is by a self-recording thermometer, of which various patterns are now available. One specimen is on the lecture room table and is being used to show the variations of temperature which take place in the course of the lecture (see Figure 17). I will subsequently put upon the screen a reproduction of the records taken in one of the most interesting of crowded buildings in London which accommodates some 5000 persons daily. First I will show,

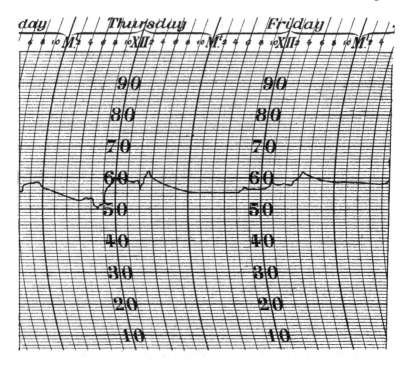

Fig. 17. Temperature changes during the lecture of Thursday, February 19, 1903 (4.30 to 5.30 p.m.).

Figures 18 and 20, the normal temperature variations in the building on a number of days which prove that, from the time of occupation of the room, the temperature increases very rapidly in consequence of the distribution of breathed air. Ultimately the tendency to rise diminishes, and the temperature falls when occupation is over. Next I put on the screen the course that the temperature ought to have followed if the building had been ideally ventilated, Figure 19, and call attention to the points of the diagram.

Finally I shall show a curious result obtained on the 23rd January 1901 which is of historic interest because that day was the day following the late Queen's death. The curves, Figure 20,

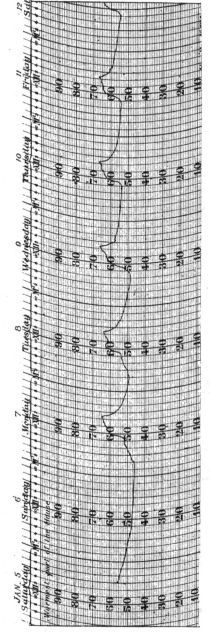

Fig. 18. Automatic record of temperature in a crowded building. January 5 to 12, 1901. The lines along which the figures are printed are the noon-lines of each day. The rise of temperature which is indicated at about 4 a.m. is due to the commencement of artificial warming for the day. Occupation commences at about 10.30 and at first the effect of many open doors keeps the temperature from rising. When the building is filled a very rapid increase of temperature of 10° is produced. Then there is almost a balance between the warming due to occupation and the cooling due to draughts. At 4.30 the building empties and the temperature falls rapidly to the night level. The rise begins again when the stokers come on duty in the morning. The building is not occupied on Sundays.

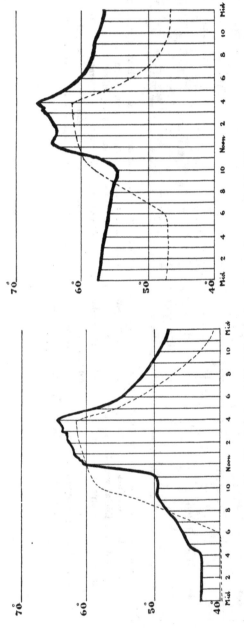

Fig. 19.　Actual and Ideal Temperature changes in a crowded building.

A cold day. Jan. 7, 1901. External temperatures
　8 A.M. 28° F. 6 P.M. 29° F. Max. 33° F. Min. 25° F.

A warm day. Jan. 17, 1901. External temperatures
　8 A.M. 46° F. 6 P.M. 45° F. Max. 49° F. Min. 34° F.

The thick continuous lines represent the actual changes of temperature on a cold and a warm day respectively, in January, 1901. The dotted lines show the ideal curves. Occupation is supposed to commence at 10.30. Following the ideal curve, the low temperature of the night is supposed to be due to widely opened windows. At 6 a.m. warmed air is supplied until by 10.30 a.m. a temperature of 58° is reached. The supply of air is thereafter so large that the presence of the persons between 10.30 and 4 only results in a rise of temperature of the house of 4° F., whereas under actual conditions the rise is about 15° on the cold day and 12° on the warm day. Under actual conditions on the cold day the temperature on entering is below 50° F. and rises by personal warming very rapidly to 61° F., then more slowly to 64°.

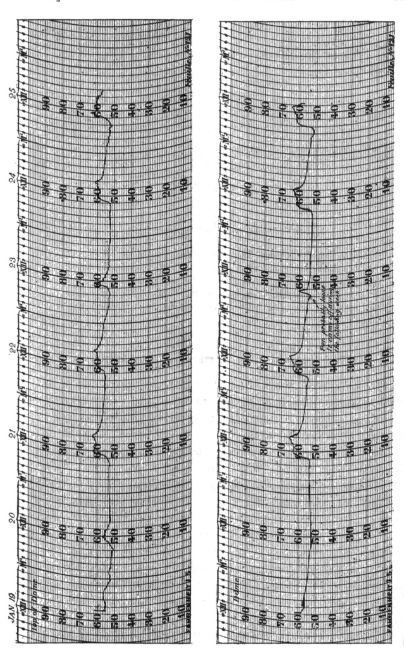

Fig. 20. Temperatures in a crowded building for the week including January 23, 1901. The lower diagram gives the temperature at about 8 feet from the floor, the upper diagram the temperatures for the same period at the top of a dome about 70 feet high. The similarity of the two curves shows how rapidly and completely the warmth is distributed by the convection currents.

show too faintly upon the general diagram. I can show it more in detail in the separate diagram which has been constructed from the records of another apparatus (Figure 21).

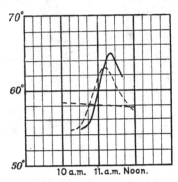

Fig. 21. Stock Exchange, January 23, 1901.
The full line shows the temperature of the air 15 feet above the floor.
The curved dotted line, the temperature of the air at the floor.
The straight dotted line, the temperature of the air in the supply channel.
The measurements were made with a Callendar electrical thermometer.

The convection of Particulate Impurity.

An additional physical principle requires consideration in view of possible effect of pathogenic organisms which may be contained on solid or liquid particles mechanically suspended in air. Such organisms may be raised in the air as dust or conveyed to the air in droplets by the act of sneezing, coughing or even energetic speaking. We have here to deal with what must be regarded as foreign particulate matter carried mechanically by the air. The conditions of carriage are obviously of importance when the removal of pernicious particles by means of ventilation is the subject of consideration.

The physical fact which comes in for consideration here is that solid or liquid particles, however small, are always falling, slowly it may be, but still falling through the air which carries them. The rate at which they fall depends upon their diameter, but no solid or liquid particle is so small or light as to actually *float* in air. It may be that the air which surrounds them is rising more rapidly than they fall through it, so that the net result is that they are carried with the current, but if the current stops they fall, and if there is stillness for a sufficient time they reach the ground. This question of the removal of particulate matter by air-currents has a very important bearing upon the methods of ventilation. It is clear that the particles left to themselves would reach the floor; a vigorous upward current might carry all with it but when it becomes less vigorous the larger and heavier particles fall faster than they rise (if that form of expression may be allowed), and only the lighter ones are carried on upward.

Indeed for every upward velocity there is a particular size of particle of water, for example, which can be just supported by the current, which sinks through the air just as fast as the current marches upwards. Heavier particles sink in such a current; lighter ones are carried upward with it. Which class of particles will be just supported and retained in position depends upon the current and the size and weight of the particles.

The law of fall of particles through air has been given by Sir G. Stokes. Of the sizes of the solid or liquid particles contained in the air of a room we know as yet practically nothing, but it is clear that what is known as downward ventilation has the property of insuring that the particulate impurity goes down with the current, whereas upward ventilation may, by its dynamical effect, be the means of supporting pernicious particles in the air.

A disturbing consideration arises from the fact that each individual himself forms a rising convection current which will carry particulate impurity, if the particles are less than a certain size depending on the strength of the upward current. The problem presented is whether it is better to bring these particles down again or to remove them, by cross ventilation or otherwise, before they have time to fall. The answer to the question really depends on the sizes of the particles; and observations of the actual rates of fall of particulate matter in still air would be of great service in this important part of the subject.

III.

APPLICATION OF PHYSICAL LAWS TO PRACTICAL VENTILATION.

WE have already seen that the supply of air to a ventilated space is subject to four laws, of which the first two, viz. the law of continuity of flow and the law of relation of aeromotive force and flow ($H=RV^2$) are independent statements. The other two, viz. the law of successive orifices, and the law of the distribution of flow between parallel orifices, are deduced from the first two in a manner similar to that by which Kirchhoff's laws are deduced from Ohm's law in the corresponding electrical problem. We have also seen that the distribution of air in the ventilated space is determined by the application of the laws of energy, momentum and thermal convection; and that the last must be regarded with the closest consideration. Further, the law of convection of fresh air must be distinguished from the law of convection of breathed air, because for the former only the temperature need be regarded, whereas for the latter the change of density due to changes in the composition must be taken into account. We have now to see how these laws may be applied in practical ventilation.

First of all we must arrive at some understanding as to what rate of air supply for each individual shall be held to constitute adequate ventilation. In considering the fundamental laws, I referred to unit flow, i.e. one cubic foot per second—very nearly 100 cubic metres, or something more than the tenth of a ton, per hour—for each person as a standard supply. That amount is very large compared with what is usually supplied in ventilation. It was computed by De

Chaumont as the amount necessary so to dilute the carbonic acid gas expired by an adult male in repose that the excess of carbonic acid gas in the diluted air, over that in the fresh air supplied, is maintained at two volumes per 10,000; that limit was further identified as the limit of perception of closeness by smell. According to De Chaumont men in active work would require more air, women and children in repose less.

From observations upon a number of schools, Carnelly, Haldane and Anderson assigned a much less amount as the minimum necessary to keep the atmosphere reasonably sweet; but, looked at from another point of view, there is a physical ground for adhering to De Chaumont's standard. As already pointed out, the air of a room is warmed by the occupants. Assuming that each occupant produces and therefore must get rid of 284 lb. F. units of heat per hour, the temperature of 3600 cubic feet supplied to him would be raised to the extent of 4° F., under ordinary circumstances, by the person who uses it. Thus, supposing a person to receive his allowance of one cubic foot per second, it will be 4° F. warmer when it leaves him than when it reaches him. This is the amount of rise of temperature that was regarded as ideal in the diagram on p. 52; any greater rise would cause stronger convection currents for the incoming air.

It is true that no allowance is made here for loss of heat through walls or windows, but for many reasons that loss should be diminished as far as possible; it is of no advantage whatever to ventilation. The amount of loss, as compared with the heat received from occupants, is very different in different buildings, being less in large crowded buildings than in small ones only partially filled. If we wish for a unit that is suitable as an ideal for all classes of buildings, although perhaps not attainable, or at least not yet attained in any, the standard unit of 1 cubic foot per second or 100 cubic metres per hour has very much to recommend it.

If this allowance is regarded as too high some other should be adopted after due deliberation. The question should not be left to be settled by what happens to be convenient or inconvenient in the structural arrangements. I have seen plans, for example, in which the allowances for two adjoining rooms used for quite similar purposes were as nearly as I could estimate them 300 cubic feet per occupant, and 3000 cubic feet, respectively. That disparity did not arise from any malevolence of design as regards one class of occupants, but merely because structural arrangements made the provision more difficult in one case than in the other. In this particular case, when I pointed out the disparity I was solemnly informed that the great advantage of the scheme was that it was practical and not theoretical, and in practice it would be found to be adequate. I have heard a similar reason given for many other practices that are not theoretically appropriate.

The next point to be considered is the temperature at which air should be admitted. Upon this point I would suggest that some approach to *uniformity* of temperature is desirable in an inhabited room, in fact the nearest approach to uniformity that is within reach. If you would have a room warm, aim at its being uniformly warm ; if cold, uniformly cold. By uniform, I mean the same in all parts of the room, not necessarily the same throughout the day. It is futile to try to keep a room up to 60° F. with an open window supplying at 28° F. the unit of ventilation for each occupant. The law of convection insists in such a case that an extremely cold layer will be found near the floor and the reversed correlative of the therapeutic action of putting one's feet in hot water will be set up. If you depend upon unwarmed air in very cold weather adopt the open air system at once and provide appropriate clothing. Uniform and draughtless cold is not disagreeable to persons suitably clad. But marked differences of temperature applied to different parts of the body, like some other drastic remedies, should only be used by special prescription of a duly qualified medical man.

Suppose however you desire to maintain the room at a temperature which may be called warm in cold weather, say between 58° F. and 62° F. The proper temperature for the entering air is 58° F. if the occupants themselves warm it through the permissible 4° F.; if there are not sufficient occupants to do this an open fire is a very comfortable appliance for the purpose, and

the entering air may be even a little colder than 58° F.
But if it is much colder the inexorable law of convection will
convert the air supply into a cold stream along the floor;
hence we cannot allow very much difference of temperature
between the air of the room and that of the air supply; convection forbids.

The next point that claims attention is one that has often
been discussed, namely, where should the air be introduced and
where should its aperture for removal be ? I shall consider the
question of removal first. To obtain some guidance in this
matter let us turn to the application of the law of convection,
with the understanding already mentioned that the convection
currents are caused by the loss of heat through windows and
walls on a cold day, and that this loss of heat is pure loss and of
no utility whatever as regards warming and ventilation. When
a room has been warmed for some time, the loss through walls
will have reached a steady state; it will not be very large and is
perhaps of some benefit as promoting a circulation of air in out
of the way corners. The main cause of loss of heat lies in the
large areas of cold window surface. Let us trace the course of
convection more closely. Suppose the air supply is at 58° F.,
the room at 62° F., the entering air will fall slowly, the breathed
air will rise in the first instance for the reason given on p. 48.
The windows will set up a vigorous downward shower, consisting
of what was originally breathed warm air now mixed with the
supply. The shower of cold air down the windows is therefore
just the air one would wish to remove and, hence, an obvious
place for putting openings for removing air is at the window
sills. The plan has two conspicuous advantages; it removes
the air which ought to be removed, and prevents it mixing
again with fresh air; it removes the coldest air and avoids the
useless waste due to loss of heat through window panes. With
some confidence, therefore, I should place a series of openings
for removing air along the window bottoms.

It may be urged that the plan omits to notice that the
chinks in the window frames are one of the sources of fresh air
supply, in some cases the most important source. That is true;
but I propose to disregard chinks as a source of air; I am

referring to rooms that are supposed to contain a number of occupants and, when the occupants exceed three or four, chinks become quite inadequate and ought no longer to be regarded. In any case the downward convection currents from windows in cold weather are an element of the problem that demands solution *. In churches and other large buildings they sometimes produce most astonishing winds. There are three other ways of dealing with them which might be adopted under special circumstances : (1) to introduce warm fresh air along the window bottoms and mix it with the descending cold shower, (2) to interpose at the foot of the windows a special source of warmth, a hot water or steam pipe, or a row of electric lamps in case the window is high up, as a clerestory window in a church, and (3) to reduce the evil somewhat by double glazing. These methods are after all only second best ; the appropriate method is to take advantage of the falling air and remove it.

As to the admission of air, it must for obvious reasons be above head height and should be on the opposite side of the room to the windows which are over the row of outlets. If there are windows on both sides the entrance openings may be between each pair of windows, and may be 8 to 10 feet up, the exits being in the window bottoms alternating with the inlet shafts. Other variants might be suggested, and might be necessary for the sake of meeting special circumstances. For the sake of simplicity of idea I propose to regard the ventilation of a room

---

* In the case of very lofty buildings with large windows, such as churches, the difficulties of warming are exceptionally great. Any departure from uniformity of temperature must necessarily produce convection currents which are sometimes very vigorous. It is perhaps an extravagance to attempt to warm such a building at all, and certainly the attempt to keep the air at a uniform temperature of 60° or 62° on a very cold day is likely to be attended with insuperable difficulties. A simple plan of securing uniformity would be not to warm the building at all, or only very slightly, enough to keep it dry, and make up for the lack of warmth by suitable clothing ; but, if something more ambitious is entered upon, attention should be largely concentrated on securing uniformity. Keeping the gas burning to warm the upper part is one way, but not a very nice one. Some attention might be given to dealing with the matter by an endeavour to warm the walls before the building is required. Probably if the walls could be warmed by energetic heating beforehand the heating machinery might be more or less cut off with advantage while the building is actually occupied.

in a generalised manner as provided for by a row of outlets on one side and of inlets on the opposite side.

Size of Openings.

The next matter for consideration is the size of the openings for admitting and removing air. This consideration enters into the problem in two ways—first as affecting the resistance of the circuit and secondly as affecting the distribution of air in the ventilated space. From both points of view the openings should be as large as possible. It is sometimes stated that there should be a definite numerical relation between the area of the inlets and that of the outlets. The basis of principle in such a relation would seem to be that because you find it exceptionally difficult to get air in, you should make it equally difficult to get it out again. As a matter of fact it may be indeed of little use to make large outlets to correspond with small inlets, because the air has got to go through both, and facility of passage in one does not compensate for difficulty in the other; but there is no objection to large outlets if the inlets are unavoidably small, but rather the reverse, provided that the outlet can be relied upon to act as an outlet; the restriction of inlets may sometimes result in the inversion of function of an outlet as in the experiment of p. 31.

The provision of adequate inlets and outlets, including the conduits for the supply and removal of air, is now the main difficulty in the way of satisfactory ventilation. No casual or deliberate defect in this respect can be remedied afterwards by increased motive power, the proportionate increase of motive power required being so large.

It may not be out of place here to remind those who have charge of building operations that the efficiency of a channel for the supply of air is determined by its narrowest and not by its widest or even its average section. I well remember the consternation with which I learned that a 9 in. × 9 in. channel intended for the supply of air to a warm-air grate in a certain building was to be contracted to 9 in. × 2 in. in order to cross a "trimmer," and the assurance which sought to allay my anxiety by pointing out that the trimmer was only 3 inches wide and the reduction of size for 3 inches would be unimportant.

As regards distribution of air the advantages of distributed orifices of entry and exit are sufficiently obvious. The distribution may be supposed to begin from the moment of entry of the air into the free space of the room and to cease at the moment of removal. To provide for the conveyance of fresh air from a narrow orifice to every part of the room, and the conveyance of used air from everywhere to a narrow orifice somewhere, is evidently a dynamical and thermal problem of the highest order of difficulty, even when objection on the ground of draught is waived. The problem is clearly simpler if air enters equally all along one side, and is removed equally all along the opposite side of a room. It becomes at once a problem in two dimensions instead of three dimensions. With sufficiently extensive and distributed orifices success may be secured without actual distribution throughout the entire length of the wall; but as an ideal we may suppose air introduced and removed along the whole length of opposite sides. It at least simplifies the diagrammatic representation of the solution. This mode of distribution may be arrived at by a distributing channel carried along the wall communicating with a tube for delivery or exhaust at one portion only. In the case of a very wide building, supply round a row of internal columns may take the place of supply along a wall.

Distribution of head.

We next consider the distribution of head necessary to produce a satisfactory supply. When the distribution takes place by means of distributing channels on opposite sides communicating with single conduits of supply and removal, the chief part of the resistance of the circuit is in the conduits; the head, or aeromotive force, is required to overcome these resistances. If the circulation is maintained by a source of power producing a head H, in order to overcome resistances r and r' of inlet and outlet respectively, it follows from Law III. that H

may be regarded as made up of parts h and h' corresponding to the inlet and outlet respectively, and that, for a flow V, $h/r = h'/r' = V^2$ where $h + h' = H$. It is on many grounds desirable to employ separate sources of power to provide the partial heads h and h' and not to use a single source for both. The grounds for this preference are as follows:

(1) If a single source of power is employed the energy required to overcome the resistance has to be transmitted pneumatically and as already pointed out (p. 38) pneumatic transmission of power is uneconomical.

(2) If the source of power is in the inlet, as in the plenum system, an excess of pressure is produced in the ventilated space and air escapes by any casual opening; on the other hand, if the source of power is in the outlet, diminution of pressure is produced in the ventilated space and air enters by every casual opening; in either case the whole working of the ventilation is disturbed by temporary or permanent variations in the resistances of parallel orifices. If there be aeromotive force in both inlet and outlet, and the resistances be suitably adjusted, the ventilated space will show no difference of pressure and the ventilation will not be disturbed by openings of doors or windows. For reasons which will be sufficiently obvious to electricians, I will give the name of "zero-potential" to the arrangement under which this state of things is produced, that is to say when the inlet and outlet are each provided with an appropriate head for the flow, and shall regard our "zeropotential" arrangement as the one which corresponds most nearly with the ideal for the purpose of obtaining a properly controlled system of ventilation. In dealing with the temperature of air to be supplied, I was careful to point out that the uniformity which I regarded as desirable is uniformity over the whole ventilated space, not uniformity from hour to hour. According to Maxwell the appropriate name for uniformity in the latter sense is constancy, and a constant temperature is not so clearly desirable. In fact the average human temperament in good health probably prefers a varying temperature, or to speak more strictly, a temperature variable at will, according to the humour of the individual at the time. There are occasions, even on a cold day,

on which the breath of fresh air to be obtained by opening windows
is acceptable to every one, except perhaps hospital patients
in certain stages of disease.   Moreover, in the variable climate
of this country, more rapid adjustment of temperature is desirable than can be supplied by the facilities for stoking a furnace.
It is therefore almost essential to preserve to the occupants of
a room the right of user of an open window, and any system of
ventilation which of necessity takes away that right, either by
reason of the draughts in the room itself, or by reason of the
interference with the other parts of a system, must occasionally
expect uncompromising criticism.

Elements required for the Ventilation of a room.

I have now indicated all the elements required for the
ventilation of a room, by means of air which can be artificially
warmed if necessary.   The elements may be summarised here:

(1) *Supply of Air*—1 unit of ventilation for each person.

(2) *Temperature*—about 4° F. below the temperature of
the room.

(3) *Orifice of Exit*—a distributive channel along the
"window wall" with openings along the window bottoms communicating with an exhaust duct.

(4) *Orifice of Entry*—a distributing channel above head
height extending along the wall opposite the windows and
communicating with a supply duct.

(5) *Motive Power*—partly in the supply duct, partly in the
removal duct, being adjusted proportionally to the resistances
of the ducts.

The intrinsic importance of this arrangement of the ventilation of a room lies in the fact that it may be regarded as
independent of the ventilation of adjoining rooms, and so form
an independent unit of a ventilation system for a whole
building.   I propose therefore to call a room considered with
reference to the supply and removal of air a ventilation cell,
and to represent it by a special diagram (Figure 22) thus:

V represents the ventilated space: numerically it represents
the flow in cubic feet per second (or hundreds of cubic metres

per hour) and, therefore, it indicates the number of persons to be accommodated in the space. The parallel dotted lines at I and O represent the distributing channels of inlet and outlet respectively.

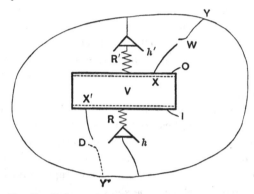

Fig. 22.   Ventilation cell with aeromotive forces and resistances
in inlet and outlet.

The zig-zag line R represents the resistance of the inlet duct; the corresponding zig-zag R' denotes the resistance of the outlet. The symbols **A** represent aeromotive forces or heads which tend to move the air. That marked h is the aeromotive force of the inlet, that marked h' the aeromotive force of the outlet; the two together, $h + h' = H$, give the whole aeromotive force of the circuit, which, in the diagram, is completed by a line YY' representing the outside air. The line XY, containing a break W in it, represents the communication with the outside by means of a window which can be either open or shut. It therefore represents a resistance, which can be varied at will between zero, when the window is wide open, and "infinity" when it is shut. A similar connexion $X'Y'$ on the other side represents a door D, the resistance of which is similarly adjustable. In this case the connexion with the external part of the circuit is dotted because the door communicates primarily with a passage, and the connexion between the passage and the external air is undefined, although often very real.

We can now lay down the conditions for efficiency in the ventilation cell V. Suppose the room is a class-room intended to accommodate 30 persons, then $V = 30$, and the flow must be 30 cubic feet per second. R and R' are determined by the construction of the ducts; $(h + h')/(R + R')$ must be equal to V^2, h and h' are determined by the means used for causing the flow. In order that W may be opened or closed without affecting the ventilation, $h/R = h'/R'$. The condition that D may be open or closed without affecting the ventilation is indeterminate, because there may be some pressure difference, or aeromotive force, between the passage with which D communicates and the outside air.

The actual size of the ventilated space V is not represented to scale, all that is indicated is the number of persons to be accommodated. This is an advantage in dealing with a building consisting of many rooms, because the characteristics of the ventilation can be treated apart from any representation of the plan—the separate cells can be shown separately for ventilation purposes, although their boundary walls may in reality be party walls.

In the result it must be admitted that the representation of a single cell is somewhat complicated.

It is not at all unlikely that the people who plan ventilation systems, and the people who use the buildings to which they are applied, would prefer something far less complicated. They are entitled to please themselves in the matter and to disregard, if they wish, all the incidental elements that affect or disturb the flow of air; but they are not at the same time entitled to complain of the failure of a system in which these incidental elements have been disregarded.

For those who are interested in tracing the conditions of successful ventilation, or in identifying the actual conditions in the various parts of an existing system, the diagrammatic representation will be found convenient. I will give one or two examples. The first, Figure 23, represents the conditions for a single room with an open fire. A represents the aeromotive force due to the chimney, which is usually of the order of from 5 to 10 pneumatic units; R_0 is the resistance of the chimney which may be about half a pneumatic unit; R is the resistance of the inlets. These last may be only chinks, the resistance of which may be estimated as of the same order as that of the chimney; or there may be special openings for the admission of

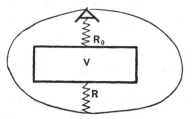

Fig. 23. Room with open fire.

air, Tobin tubes or open windows, which will diminish the resistance possibly to such an extent that the resistance of the inlets is altogether negligible.   Supposing, for the sake of calculation, that the resistance of inlets is equal to that of the chimney; the combined resistance of inlets and outlets is $R_0 + R$, i.e. 1 unit, and with aeromotive force at 10 units (Law I.) $V^2 = H/R$, gives us $\sqrt{10}$ for V; i.e. the number of persons supplied by the system, on the liberal scale of a unit per person, is three.   In the diagram the door which may connect the cell with some other cell, and therefore introduce other aeromotive forces, and also the possible effect of wind, have been disregarded.

The next case for illustration is two adjoining rooms, each with open fires and inlets.   The diagram, Figure 24, shows two

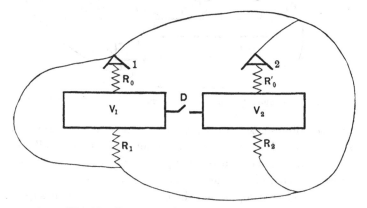

Fig. 24. Two adjoining rooms, each with a fire.

cells, and the possibility of connexion between them shows how the aeromotive force in one may affect the flow in the other. This case is really similar to the analogue of the potentiometer already described p. 30, and the equations derived from the pneumatic laws would enable us, by knowing or assuming the values of the various quantities involved, to calculate the effect of the two aeromotive forces acting together. A_1, A_2 are the aeromotive forces due to the chimneys, R_0, R_0' their resistances, R_1, R_2 the resistances of the inlets; the variable communication between the rooms is indicated by the broken line D, connecting the two cells. V_1, V_2 are the respective flows. A simple deduction from the diagram is that the condition for "no flow" between the two rooms is that there should be no aeromotive force across D. The analytical expression for this condition is clearly

$$\frac{R_1}{R_0 + R_1} A_1 = \frac{R_2}{R_0' + R_2} A_2.$$

Figure 25 represents a still more complicated system consisting of 5 cells forming together part of the ground floor of

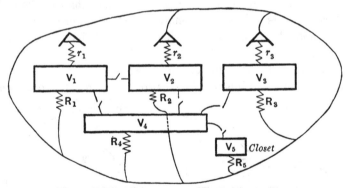

Fig. 25. The ground-floor of a house including three rooms with fires.

a house represented in Stevenson and Murphy's *Hygiene*, vol. I. p. 113. It represents three adjoining rooms V_1, V_2, V_3, each with a fire, and a passage V_4 with which the rooms communicate. There is besides a closet V_5 which, as mentioned in the article referred to, formed the chief source of the air supply. It will be evident from the figure that if the resistances

R_1, R_2, R_3, R_4 are great compared with R_5 that the flow V_5 will be the main flow. The application of the pneumatic laws to this case, just in the same way as the application of Kirchhoff's laws to the corresponding electrical problem, would furnish a series of simultaneous equations which would give the numerical values of one series of variables if those of the others were known.

In a building with many rooms the diagram would obviously become very complicated, and especially so if more than one floor had to be considered, and aeromotive forces due to convection up staircases etc. came into play. But the complication of the diagram is not greater than that of the actual circulation of air currents in the building.

I propose now to consider the application of this mode of representation, and of the principles upon which it is based, to the various systems of ventilation that are to be found in practice. In doing so I shall ask to be allowed to digress for the purpose of pointing out some interesting characteristics of the various forms of motive power which are employed in ventilation.

Open fires.

The first and probably the most common system is that of the open fire, of which the diagram has been already given Figure 23. I refer to it again only as an excuse for a digression upon the motive power in that case. The total resistance of a circuit in which the motive power is that of the hot air in a chimney is made up of that of the chimney R_0 and that of the inlet R. The flow is given by the equation

$$\frac{H}{R_0 + R} = V^2.$$

Clearly the resistance of the circuit can never be less than R_0, and it cannot be greater than the resistance of the chimney together with that of the chinks, $R_0 + R_c$. Thus it can only lie between the limits R_0 and $R_0 + R_c$. We have already seen that the numerical values of these quantities may be about the same and about 0·5 unit each. We can therefore trace the possible variations of flow for resistances between those limits, viz. 0·5 and unity.

In the first place we will suppose the aeromotive force constant, in which case we get

$$V^2(R_0 + R) = \text{constant.}$$

A curve *ABC*, representing this relation, is shown in Figure 26, where *ON* represents unit resistance, *OM* half a unit. It is

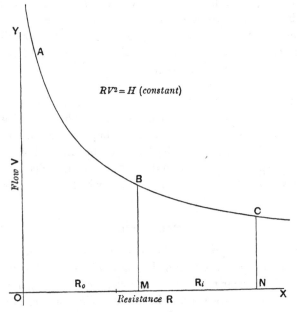

Fig. 26.   Flow through a circuit with constant aeromotive force
and varying resistance.

OM resistance of chimney.
MN   ,,   chinks.

apparent that in the case of a chimney, where R_0 is the minimum resistance and $R_0 + R_c$ the maximum, the flow only varies from *BM* to *CN* by varying the resistance, so that there is very little latitude in the ventilation accommodation of a room under those conditions.   If the room is "better built" and the chinks are less the curve would be more prolonged in the direction of *X*.

But actually with an open fire the head does not remain

constant. An increased flow of air up the chimney diminishes
the temperature and consequently the aeromotive force. If
we assume that the combustion of coal goes on at a constant
rate, we may assume that $H = \frac{\mu}{V}$ and thus $\frac{\mu}{V^3} = R$. Figure 27
shows the relation between flow and resistance under these

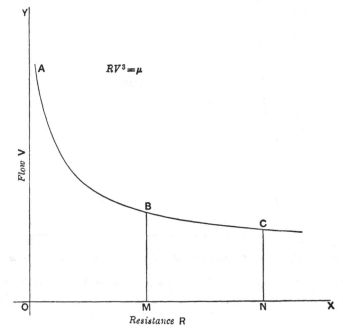

Fig. 27.   Flow through a chimney with constant consumption of coal.
OM resistance of chimney.
MN ,,   chinks.

circumstances. It shows that the variation of the flow in consequence of the variation of resistance is even less than on the
hypothesis of constant aeromotive force; and, as a matter of fact,
unless the resistance of the chinks is very great, the diminution
of inlet-resistance, as by the opening of a Tobin tube in a room
which requires ventilation, produces greater effect by its local
distribution of cold air than by its increase of the flow.

One other diagram, Figure 28, may be referred to with
reference to open fires. It represents the relation of coal
consumption and flow for a fixed resistance. It is quite

sufficient to demonstrate that the increased consumption of coal necessary to increase the flow is very great, and that the

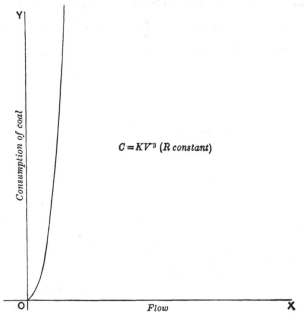

$$C = KV^3 \ (R \ constant)$$

Fig. 28.   Open fire.   Consumption of coal and flow with constant resistance.

attempt to improve ventilation by increasing the motive power is not worth the expense, at all events in the case of ventilation by open fires.

Automatic Cross Ventilation.

The next system to be considered is that of natural or automatic ventilation of rooms without fires by means of open windows or Tobin tubes.   For rooms occupied by a large number of persons, if we leave out of account the case of windows wide open when the room becomes simply a shelter, it can only be carried out satisfactorily if there are windows on both sides and the replacement of air takes place by cross ventilation. The diagram representing this case is shown in Figure 29. I have given a full discussion of the case in the Report to the Local Government Board on the Ventilation and Warming of Poor Law Schools, 1897.

The chief characteristic of this system is that the head which depends almost exclusively upon the wind is very variable and may vary in sign as well as magnitude: it is therefore marked (variable + or −) in the diagram. Consequently the

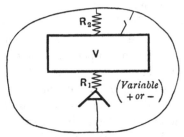

Fig. 29. Cross ventilation by wind.

system can only be expected to work satisfactorily "on the average," which is unfortunate. Assuming however an average wind of six miles per hour, the area of opening required on each side for each person to give the full amount of air is 48 square inches, which has a resistance of about 0·4 pneumatic unit. The aeromotive force corresponding with the average wind of six miles per hour is about 1·4 pneumatic units.

Automatic Ventilation by Cowls.

We may now consider briefly the diagrammatic representations of other systems. Figure 30 shows the arrangements for

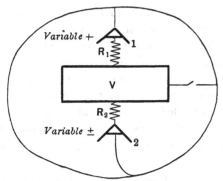

Fig. 30. "Natural" (Automatic) Ventilation by Cowls.

A_1 aeromotive force due to cowls (variable +).

A_2 „ „ „ wind (variable + or −).

ventilation by cowls and Tobin tubes or windows. It is a modification of the cross ventilation system shown in Figure 29.
Since the aeromotive force due to a cowl depends upon the
wind, except in so far as temperature differences in the shafts
which carry cowls cause aeromotive forces therein, I have taken
the wind into account in the diagrams as a separate force. I
have marked A_1, the aeromotive force due to the cowls as
variable, because it depends upon the force of the wind in
a manner set out in detail in the report of the Cowl Committee
already referred to (p. 23); but I have marked it +, as the cowl
is generally so exposed that the flow due to it is in the right
direction. The aeromotive force due to wind, apart from its
action on the cowl, I have marked variable ±, because the wind
may be on one side of the building or the other, and thus the
automatic ventilation due to the wind on the openings may be
in the same direction as, or opposed to, that due to the aeromotive
force of the cowls according to the position of the inlets and the
direction of the wind.

The Vacuum System.

Figure 31 represents the Vacuum system as applied to a
combination of four cells. There is a single aeromotive force **A**
due to a fan which has an internal resistance R_0. In the
diagram the fan is represented as drawing air through R_1, R_2,
the resistances of the outlet ducts from two of the cells, the
flows being V_1 and V_2 respectively. The two cells are provided
with special inlet ducts with resistances r_1 and r_2; but it may be
noted that there are doors d_1, d_2, d_4 which may be shut; or they
may be open, and make communication with practically no
resistance between the cells and the passage V_4, or the cell V_3
which is not intended to be included in the system. The
opening of one of these doors places the aeromotive force in
communication with the inlet r_3 or r_4, which is thus brought
into the system.

In computing the resistances r_1 and r_2 the effect of chinks
ought to be taken into account and not merely the gratings or
ducts which are put in for the special purposes of ventilation,

and besides these openings we must allow for a window W in any of the cells, which may be open or shut.

The process of ventilation in this case is the following: the fan reduces the pressure in the ducts with which it is in

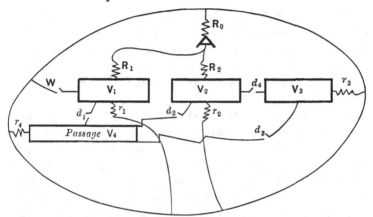

Fig. 31. 4-cell Vacuum System.
A motive power acting on the two cells V_1, V_2.
R_0 resistance of fan.
R_1, R_2 resistances of outlets.
r_1, r_2, r_3, r_4 resistances of inlets.
d_1, d_2, d_3, d_4 communication doors.
W window opening.

communication, and consequently in the cells V_1, V_2; the reduced pressure there determines a flow V_1 or V_2 into the cells to supply the outlet ducts. The distribution of flow between the two cells depends upon the resistances R_1, R_2, r_1, r_2 and any variation in any one of them means variation in the whole system. Thus suppose all doors and windows are shut, and all chinks accounted for in r_1 and r_2, then the total resistance of cell V_1 is $R_1 + r_1$, of cell V_2, $R_2 + r_2$. By Law IV. the equivalent resistance of both is R where

$$\frac{1}{\sqrt{R}} = \frac{1}{\sqrt{R_1 + r_1}} + \frac{1}{\sqrt{R_2 + r_2}},$$

and the total flow $V_1 + V_2$ due to the aeromotive force of the fan is given by

$$A/(R_0 + R).$$

The distribution between the alternative channels is inversely proportional to the square roots of the resistances, and therefore

$$\frac{V_1}{V_2} = \frac{\sqrt{R_2 + r_2}}{\sqrt{R_1 + r_1}}.$$

Hence the flow in any part is determined by the magnitude of the resistances in all the various parts of the circuit, and is subject to change when any resistance is changed. Thus suppose the window W opened, we practically reduce the inlet resistance of that cell to zero and must write $r_1 = 0$. V_1 is in consequence increased but V_2 is diminished.

In ventilation on this system the warming is sometimes made to depend upon drawing air through special inlets in which, or in front of which, heating surfaces are placed. It will be noticed that the power for producing a flow through these special inlets has to be transmitted across the ventilated space by diminishing the air pressure. But, as I have already intimated, the inlet resistance r_1 must take account of the chinks as well as the special inlet openings, and sometimes the chink resistance is small compared with that of the special inlets, so that the flow into the room is practically all through chinks and little or none through the special openings. Moreover with lapse of time the chinks increase in size, and therefore diminish the resistance, and the transmission of pneumatic power to the special inlets becomes increasingly defective. An arrangement of this character is never very easy to manage and may become very defective as a means of securing a supply of air through channels originally designed to act as inlets.

The Plenum System.

The Plenum system represented for the same four cells in Figure 32 is similar to the Vacuum system with the exception that the aeromotive force is reversed and air is blown through the inlets into the cells. It finds its way out through the channels designed as outlets and through chinks as well, if there are any, the distribution between the two being governed, in accordance with Law IV., by the relative magnitudes of the resistances of the outlet channel and the chinks. As the fan communicates directly with the inlet channels there is not the

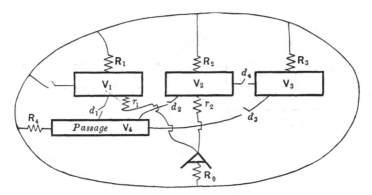

Fig. 32.   4-cell Plenum System.

A motive power acting on the two cells V_1, V_2.
R_0 resistance of fan.
r_1, r_2 resistances of inlets.
R_1, R_2, R_3, R_4 resistances of outlets.
d_1, d_2, d_3, d_4 communication doors.

same difficulty about securing the passage of the air through inlets as there is with the vacuum system; but the disturbance of the distribution of flow in consequence of the chinks, or of open doors or windows, still remains, and, for either system to fulfil its intentions, the chinks must be reduced to a negligible quantity and the doors and windows must be regulated according to the requirements of the system, and not according to the individual tastes of the occupants of the cells.

The Zero Potential System.

Figure 33 represents the zero potential system which, so far as I know, is most closely represented in actual practice by the Lyon-Cadett system of cross ventilation. The ventilation of the House of Commons is another example; but in that case the air enters through the floor from the inlet fan and passes out through the ceiling to the outlet fan which, by a recent alteration replaces the original outlet system of immense chimney shafts in two of the towers. As will be seen by the figure, there are separate aeromotive forces for the inlets and

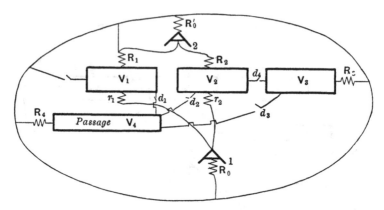

Fig. 33.   4-cell "zero potential" system.

A_1 motive power acting on inlets to two cells V_1, V_2.
A_2 motive power acting on outlets.
R_u, R_0' resistances of fans.
r_1, r_2 resistances of inlets of the cells.
R_1, R_2 resistances of outlets of the cells.
R_3, R_4 outlet or inlet resistances of cells V_3, V_4 which are without motive power
    if r_1, r_2, R_1, R_2 are properly adjusted.

outlets, and if the resistances are appropriately adjusted, the cells exhibit the same pressure as the outside air; consequently opening or shutting doors or windows makes no difference to the flow except in so far as the wind may exert some influence as a separate and independent motive power.

In the diagrams for "natural" or automatic ventilation I have represented the aeromotive force which is due to wind, because the wind is the agent upon which the systems depend; but I have omitted any representation of its aeromotive force in the diagram of the systems depending upon mechanical power, because it only affects those systems as a disturbing cause. The thermal aeromotive force of differences of temperature is also a disturbing cause. In a mechanical system steps may have to be taken to protect the ventilation against the effect of wind, and in a high building the thermal forces may become important.

Characteristic Curves for mechanical ventilation.

The consideration of the working of the various mechanical systems may be conducted most easily by reference to the diagram, Figure 34, which represents what may be called charac-

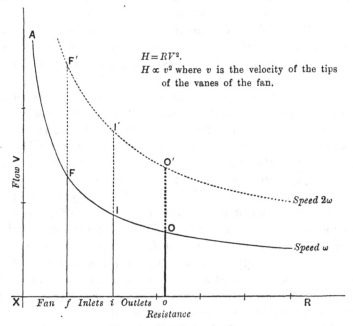

$H = RV^2$.

$H \propto v^2$ where v is the velocity of the tips of the vanes of the fan.

Fig. 34. Ventilation by fan.

$AFIO$ curve of relation of Flow and Resistance, speed ω.

Xf internal resistance of fan.

fi resistance of inlets.

io resistance of outlets.

Oo the flow through the circuit for the specified resistances with speed ω.

$O'o$ the flow through the circuit for the same resistances with speed 2ω, or with a fan of larger diameter in the ratio of $\sqrt{2}:1$ and of the same internal resistance.

Note. The fraction of the flow Oo which passes through a single outlet duct whose resistance is R_1 is (by Law 4)

$$\frac{\sqrt{\dfrac{1}{R_1}}}{\Sigma\left(\sqrt{\dfrac{1}{R}}\right)},$$

where

$$\Sigma\left(\sqrt{\dfrac{1}{R}}\right) = \sqrt{\dfrac{1}{R_1}} + \sqrt{\dfrac{1}{R_2}} + \sqrt{\dfrac{1}{R_3}} + \&c.$$

teristic curves for a mechanical system, and in that respect are
analogous to the characteristic curves for a dynamo machine.
According to M. Murgue *, a fan produces a definite aeromotive
force depending on the square of the velocity of the tips of its
vanes, and therefore of its speed of rotation. At the same
time the fan has a definite pneumatic resistance, depending
upon its area and the obstruction which it offers to the passage
of air. This resistance is inseparable from the action of the
fan, and has to be added to the resistance of inlets and outlets,
which are also additive if the cells are large compared with the
inlet and outlet channels.

The characteristic curves which I have drawn are ideal ones
with assumed values for the resistance of the fan, the inlets and
the outlets. The curve FIO represents the characteristic for
a speed ω. Resistances are measured along XR and the flow in
the circuit is the ordinate Oo drawn through the point o to
meet the characteristic curve. Xo represents the total resistance
of the circuit made up of Xf, that of the force, fi that of the
inlets, and io that of the outlets; $F'I'O'$ is the characteristic
for the double speed 2ω, and is obtained by simply doubling the
ordinates because the aeromotive force is proportional to the
square of the velocity of the tips of the vanes, and on the other
hand, for constant resistance, is proportional to the square of
the flow. Hence with constant resistance and varying speed
the flow is simply proportional to the speed.

These ideal characteristic curves are perhaps approximately
correct, but it would certainly be much better to have them
drawn from actual experiment just as in the case of a dynamo
machine†. It is usual to specify the performance of a fan by
the flow which it produces when it is mounted "on a screen,"
that is to say, when the flow has no resistances to overcome
except the internal resistance of the fan. A much more instructive specification for a fan would be its characteristic
when the flow has to pass through different known resistances.
It would not be a difficult matter to determine the characteristic
by experiment, though a good deal of space would be required

* *Theories and practice of centrifugal ventilating machines.*

† In actual practice the resistance of the fan is probably not simply an
addition to the resistance of the duct.

for the apparatus. In the absence of any such measurements we have to be content with the ideal characteristics derived from M. Murgue's generalisation.

In the diagram I have taken the resistance of the fan Xf as being equal to the equivalent resistance of the inlets fi; that of the outlets io is somewhat larger than either. This probably underestimates the resistance of the fan, which is often much greater than that of the rest of the circuit; but, for the purpose of illustration, the assumptions which I have made will serve. Any other distribution of resistances would simply mean drawing ordinates at the appropriate places to the same characteristic curve. Suppose for example we are working with the vacuum system, and the windows are opened so that the resistance of the inlets is zero. Then fi is zero and the point o is nearer X by the distance fi. The flow is therefore increased to that represented by the ordinate through the appropriate point. Or on the other hand, suppose we are working with the plenum system and the outlets are so enlarged that their resistance is practically zero, then io is reduced to zero and the flow is represented by the ordinate Ii.

Again, suppose we use a larger fan of similar construction, say one of double diameter but driven with the same number of revolutions as the original one. We now move to the characteristic curve $F'I'O'$, because the velocity of the tips of the vanes is doubled and the aeromotive force is proportional to the square of that velocity, and the flow is increased in simple proportion with the speed of the fan. But since the aperture of the fan is doubled in diameter its resistance is now much smaller, probably about one-sixteenth of that of the smaller fan; for resistances are inversely proportional to the squares of the areas of similar apertures. Consequently the resistance length Xf will be reduced to one-sixteenth. To it must be added the lengths fi and io if the inlets and outlets are unaltered, but we could now almost neglect the resistance of the fan, and to determine the flow through the circuit we should draw Xo' equal to fo and through o' draw the ordinate. In this particular diagram (since $Xf = fi$) it would be the same as the flow for the first fan with the inlet resistance zero.

s.   6

In a similar manner the flow through the circuit for any arrangement of fan and resistances can be determined.

Suppose a second fan is interposed as in the zero potential system. We now get an aeromotive force equal to the sum of that produced by each fan, and at the same time we add the resistance of the new fan. Suppose the two fans are equal and similarly driven, we double the aeromotive force, and if the resistances were unaltered, we should increase the flow for any combination of resistances in the ratio of $\sqrt{2}:1$, because the flow is proportional to the square root of the aeromotive force ($H = RV^2$). Thus we move on to a characteristic got by increasing each of the ordinates of FIO in the ratio $\sqrt{2}:1$. Then we set out along Xo the total resistance of the two fans, the inlets and outlets, and draw the corresponding ordinate to the new characteristic.

Effect of wind.

The wind blowing across or upon the openings of a ventilation system alters the aeromotive force without altering the resistance of the system. It increases the aeromotive force if it flows across the outlet, or upon the inlet, and diminishes if it blows across the inlet, or upon the outlet. The increase or diminution of head would change the characteristic from the curve of undisturbed flow to one corresponding with increased or diminished aeromotive force as the case might be. As the wind is variable the change of aeromotive force will vary in correspondence with the wind, and thus the characteristic which determines the flow will lie between two extremes, and that applicable at any particular moment will lie somewhere between the limits. The limits between which the variation may take place could easily be determined by measuring the flow in any one of the cells when wind is blowing and when no wind is blowing. The effect of wind upon the diagram is therefore to replace the line of the characteristic curve by a more or less broad band, bounded by the extreme characteristics. Between these extremities the characteristic for any special state of the wind will lie.

Comparison of Systems of Mechanical Ventilation.

Let us now compare the three systems of mechanical ventilation with the aid of the characteristic curve. For this purpose we must divide up the total flow through the system, as represented by the ordinate Oo, into the portions corresponding with each cell. We may practically assume that the flow through each cell is separate from that of its neighbours. The assumption is practically true when doors and windows are shut, and approximately true when doors of communication between cells are open if each cell is directly connected with the fan. The resistances may not be adjusted so as to secure perfect independence of flow through the separate cells, but the adjustment will probably be nearly satisfactory in that respect.

To find the distribution we must find the total resistance of each cell by adding the resistance of inlet and outlet. To do this strictly would require each resistance to be separately evaluated and the result would be very complicated. For the purpose of illustration we may assume the inlet resistance of each cell to be the same, and the outlet resistance to be the same as the inlet, and each equal to R. Thus the resistance in the circuit due to each cell may be taken as $2R$. Then since the total flow as represented by Oo is divided between each cell in accordance with Law IV., and the fraction of the flow through any cell is

$$\frac{\sqrt{\dfrac{1}{2R}}}{\Sigma\sqrt{\dfrac{1}{2R}}}.$$

For example, if there are 5 cells with equal resistance the flow through each will be $\frac{1}{5}V$.

With the vacuum system the inlets will be partly authorised ducts, probably leading over heated pipes to warm the air, and partly chinks. If the resistances of these two divisions of the inlet are equal, one half of the air supply for the cell will go through the authorised inlets; the other half will go through the chinks, and thus escape the desired warming. I am not able to assign a numerical value for the ratio of the resistance of the chinks to that of the authorised inlets; but, judging from the effect of an open fire, the resistance of chinks, unless special precautions have been taken, is probably not greater than that of the authorised inlets, and one half the air will get in through them.

Next suppose a window opened in one of the 5 cells, reducing the outlet resistance of that cell to zero. That will reduce the total resistance of that cell from $2R$ to R. The resistance of the whole circuit will be slightly diminished, and the flow ordinate of the characteristic removed a little nearer X, since $\Sigma \dfrac{1}{R}$ is a little greater (see Law IV.), but this effect when there are 5 cells is unimportant. The important effect of short circuiting the outlet of one cell is upon the distribution of the flow. The cell with the open window gets twice as much as any one of the others and thus gets two-sixths of the whole flow, whereas each of the others gets one-sixth instead of one-fifth as before. Thus the opening of the one window produces a considerable readjustment, and in addition to the advantage of an open window the one room gets a doubled flow, at the expense of the other cells, which will appeal to different persons as an advantage or a disadvantage according to temperament and other circumstances. If in addition a door is opened from the cell with an open window to a neighbouring cell, two outlet resistances are short circuited, and the disturbance of the flow is still greater. The same is true of the plenum system, which is however free from the disadvantage of the uncertainty of inlets.

With the zero potential system the motive power is supplied for the inlets and outlets separately; and, supposing they are the same in magnitude, the characteristic curve would be the portion FI of Figure 33 duplicated. In that case two fans would be used of the same dimensions as the one for which the characteristic is drawn in Figure 33 and the flow would be Ii instead of Oo (supposing the inlet and outlet resistances were equal—in the figure they are not so represented), but in that case the two fans would involve the expenditure of double power. To get the same flow as with the single fan the fans

could be used at lower power, but the fact of the addition of
resistance in the fan itself would mean some expenditure of
power greater than is necessary for the same flow in a current
with a single fan. The advantage would be that the flow would
take place through the inlets and outlets, as designed; the
chinks in the cells would not be called upon either to supply or
receive air, and the opening of doors and windows would produce
no interference with the action of the fans.

Numerical computations.

In conclusion I should like to give some idea of a method of
computing the dimensions of fans and ducts necessary for the
supply of air on a mechanical system. Take for example a
36 inch fan of low pressure type, to ventilate 6 rooms. Suppose
each room has an inlet duct 28 in. by 9 in. and an outlet duct of
similar dimensions. By making a number of assumptions we
can form an idea of the volume of supply; we found in p. 22 that
the resistance of a chimney about 20 ft. high, and 14 in. × 9 in.
in section was about half a unit. The resistance of a duct
is inversely proportional to the square of the area. The resistance of a duct does not depend to any great extent on its
length, if added length does not involve addition to the number
of bends or obstructions. Thus we may estimate the resistance
of an inlet duct of twice the area of the chimney at $\frac{1}{8}$ unit. The
resistance of inlet and outlet for each cell is therefore $\frac{1}{4}$ unit.
The 6 cells in parallel will have an equivalent resistance $\bar{\bar{R}}$,
determined by Law IV.

$$\frac{1}{\sqrt{\bar{\bar{R}}}} = \Sigma \frac{1}{\sqrt{R}} = \frac{6}{\sqrt{\frac{1}{4}R}}$$
$$= 12.$$
$$\therefore \ \bar{R} = \tfrac{1}{144} \text{ unit.}$$

The fan having a diameter of 36 inches, and the aperture
of the standard unit resistance a diameter of 6 inches, the
resistance of the fan will be $(\frac{1}{6})^4$, i.e. $\frac{1}{1300}$ approximately.

Thus the total resistance in the circuit is $\frac{1}{1300} + \frac{1}{144}$ units,
or approximately $\dfrac{1 \cdot 1}{144}$.

A fan of 36 inches, revolving at 120 revolutions a minute, gives, at 20 ft. per second for the tips of the vanes, about 1·34 units of aeromotive force (see Table, p. 9), and hence the flow would be

$$\frac{1\cdot34 \times 144}{1\cdot1} \text{ or } 175.$$

Such an arrangement would thus give an ample supply of air for 175 people.

In a similar manner by introducing appropriate values for the quantities involved the numerical relations of any other problem can be determined.

Conclusion.

I have now finished my exposition of the physical principles involved in the various systems adopted for the ventilation of buildings; and I trust that I have made it clear that it is possible to regard the behaviour of air currents as subject to perfectly definite laws, although it may be difficult or even impossible to introduce accurate numerical values for all the different elements which are involved in the determination of the magnitude or behaviour of all the various currents which make up a system of ventilation. It will thus be possible at least to think clearly of the problems to be faced, and the closer investigation of the laws applicable to air currents in the process of ventilation is not unworthy of the attention of students of physics.

I have referred more than once to the complication of the problems presented, but I do not think that need deter us. I do not find that those who have most objection to the complexity involved in the statement of the problems of ventilation, when they are reduced to their lowest terms, have any hesitation about asking for advice in such a form as to present a problem of doubled complexity. I have often been asked to suggest a remedy for cases in which a ventilation system fails to give satisfaction, and the answer desired is generally the mention of some simple appliance which may be superposed upon the original system. Consider what that means. For example, a building is ventilated upon the open fire system and

the provision is inadequate and a remedy is sought. The
extension of the original system is out of the question, we
cannot practically increase the currents through the chimneys
for the reason given on p. 72; we cannot build more chimneys,
partly for structural reasons, and partly because the distribution
of heating would become hopelessly disarranged. So the
ingenuous inquirer would like another system superposed.
That means combining two of the diagrams which I have
brought before you, but if one of them seems too complicated
for practical purposes what is to be said of the combination of
two? It is difficult enough to think out, still more difficult to
introduce in practice with success. I can see no solution for
the problem of remedying a system which is found to be
insufficient, except to bring it into line with one or other of the
systems mentioned by modifying some portion of the circuit.
The combination of two might be accomplished by a patient
experimentalist by successive trials, but the specification of the
supplementary arrangements à priori would surely dismay the
most enterprising enthusiast with a real knowledge of the
subject.

When all is said and done, the successful management of a
system of ventilation is a physical experiment of great difficulty.
The interference which is caused by the wind and other conditions
of weather cannot be disposed of satisfactorily without more care
and intelligence than is necessary to mind a fire. We know
now almost enough of the behaviour of systems of ventilation
which fall short of being completely successful, and what is
wanted for future guidance is an experiment, not on the gigantic
scale of a new collegiate building, or even of the debating
chamber and lobbies of the House of Commons, but on the
small scale of two or three good sized rooms, by an installation
for the purposes of experimental research, in which there is
ample margin in the motive power, and ample elasticity in
the regulation of inlets and outlets and in the distribution of
temperature—a true experimental installation. With such an
installation we might be bold enough to hope that the conditions
of really successful ventilation might be exemplified, and the
effect of restrictions on one or other element might be examined.

The provisional outlay would doubtless be considerable, but it need not be very large; in the long run it would, I feel sure, result in an enormous saving of money as well as a substantial increase of comfort.

To introduce a system of ventilation under restrictive conditions as to the supply of air is not to solve the physical problem of the adequate ventilation of a building but to attempt an answer to the more difficult question—if the conditions of success be interfered with by this or that arbitrary modification, will the resulting state of affairs satisfy those who occupy the rooms? a question which has indeed but little interest from the point of view of practical physics.

NOTES.

Note A. Measurement of the flow of air.

THE commonest form of apparatus for measuring the flow of air in ventilation circuits is an air-meter. It consists of a light wheel with a number of radial arms which carry blades set at an angle of 45°. It can be obtained from any instrument maker.

In the more usual form of the instrument the revolutions of the wheel when exposed to the air current are counted by a train of light wheel-work which moves a series of pointers over counting dials. The dials are graduated to give the total of the air flow in feet, for the time during which the instrument has been exposed to the current. A small correction, amounting usually to 30 feet for each minute's run, has to be added to the reading derived from the dials. A full account of experiments to test the readings of an instrument of this kind, as used to determine the flow through a pipe of about the same diameter as the revolving wheel, is given in the Report of the Sanitary Institute upon the work of the Cowl Committee to which reference has already been made.

Such instruments are also tested at the National Physical Laboratory (Kew Observatory) by means of a whirling table. The results of these tests give the readings of the meter when it is exposed in what may be called the free air in contradistinction to the air confined by a tube.

Notice should be taken of the fact that these two methods of using the meter, viz. in a pipe of the same diameter as the instrument and in the free air, must be expected to give different readings. In an exposure of the first kind, i.e. in a tube, what has to be measured is the actual flow through the tube; the instrument itself offers some obstruction and reduces the flow. In the free air the reading of the meter is also affected by the obstruction offered by the wheel. The measure aimed at, however, is not that of the obstructed flow in the limited portion occupied by the wheel, but the practically unobstructed flow in the larger space which surrounds it. In this case the corrected graduation of the instrument makes allowance for the obstruction, while in the former the obstruction is in the flow which is to be measured.

In making a measurement with the air-meter the plane of the revolving wheel should be set at right angles to the current to be measured. Moreover, in order to determine the flow through a grating or opening of considerable size regard must be had to the distribution of the current over the opening. Through some openings the flow will be found to vary very considerably in different parts; it may even be restricted to a few inches at the top or bottom. A general instruction is to make allowance for the variation in different parts as the circumstances of any particular case may indicate.

A modification of the instrument consists in controlling the motion of the free wheel by a spring. Thus the reading of the deflexions of a pointer against the force of the spring is substituted for the counting of the revolutions of the wheel as the index of the strength of the current. The advantage of this statical method of estimating the current as compared with the original form of the air-meter is that the time of exposure does not enter into the measurement. Some years ago my assistant, Mr A. G. Bennett, made for me a number of instruments with a similar purpose. They consisted of a short tube of square section with a mica flap at one end. They could be easily hooked to a ventilation grid. The mica flap was weighted, and its deflexions were so graduated as to give by direct reading the number of units of ventilation for a square foot of the grating which carried the current shown by the apparatus.

Either of these forms of statical apparatus illustrates the difference in the method of using an air-meter to which attention is called in this note. The obstruction offered by a standing wheel or a mica flap is much more evident though not more real than the obstruction offered by the revolving wheel.

Note B. Data for the determination of the head or aeromotive force due to various agencies. See Table p. 9.

The columns of the table given on p. 9 express the results of a number of calculations from experimental data, which enable us to compute the head or aeromotive force produced under various conditions.

The first four columns merely represent alternative methods of specifying the head. The fifth column giving the wind required to produce "head" by direct impact is computed on the basis of the experimental results of Mr W. H. Dines, F.R.S.*, from which it appears that the relation of the wind velocity in miles per hour to the pressure, in pounds weight per square feet, exerted by it upon a circular disc one square foot in area is given by the equation

$$P = \cdot 003 \ V^2.$$

The figures in this table differ from those of the corresponding column in the table in Stevenson and Murphy's *Hygiene* in consequence of the use of Mr Dines's factor ·003 instead of the factor ·005 previously employed.

The sixth column giving the wind required to produce "head" by the "suction" due to cross-blowing is derived from the formula $H = \dfrac{v^2}{2g}$ given in Stevenson and Murphy's *Hygiene*, vol. I. p. 76. It was roughly confirmed by experiments referred to on the page indicated.

All wind measurements are still matters of considerable uncertainty and the practical application of such results as are given in the table depends largely upon details of exposure. No stress should be laid upon the difference in the figures of the two wind columns.

* See Report on the Beaufort Scale of Wind Force, Meteorological Office Publication, No. 180, p. 6, 1906.

The figures for the centrifugal and helicoidal fans in the seventh and eighth columns are taken from M. Murgue's book on centrifugal ventilating machines. Those for the ninth and last column are computed from the coefficient of thermal expansion of air, the head being taken as $\dfrac{\rho - \rho'}{\rho'}\, h$, where ρ, ρ' represent the density of the air outside and inside the chimney respectively; h is the height of the chimney.

Knowing experimentally the relation of the density to the temperature we get the head, H equal to

$$\frac{T-t}{459+t}\, h,$$

where T is the temperature of the air in the flue and t that of the external air. From this formula the temperature necessary for the specified values of the head are calculated.

INDEX.

Printed in the United States
By Bookmasters